STATISTICA

Visual Basic Primer

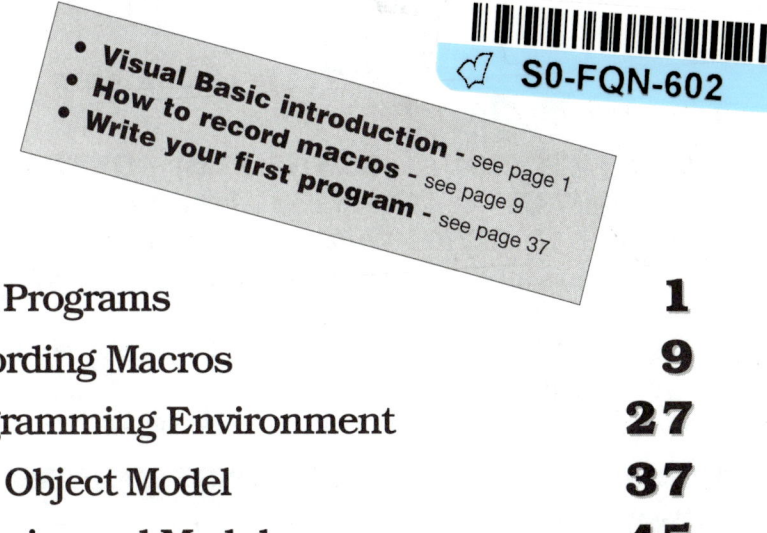

- Visual Basic introduction - see page 1
- How to record macros - see page 9
- Write your first program - see page 37

Chapter

1:	SVB Programs	**1**
2:	Recording Macros	**9**
3:	Programming Environment	**27**
4:	SVB Object Model	**37**
5:	Libraries and Modules	**45**
6:	Recorded SVB Programs	**53**
7:	Customizing *STATISTICA* with SVB	**65**
8:	Matrix and Statistical Functions	**81**
9:	Introductory Examples	**87**
10:	Advanced Examples	**95**

Appendix

A:	Events: Document-Level	**125**
B:	Events: Application-Level	**133**

www.statsoft.com

STATISTICA VISUAL BASIC PRIMER

Table of Contents

1. *STATISTICA* VISUAL BASIC (SVB) – OVERVIEW 1
    Creating *STATISTICA* Visual Basic Programs 3
    Executing *STATISTICA* Visual Basic Programs 4
    Applications for *STATISTICA* Visual Basic Programs 5
    Structure of *STATISTICA* Visual Basic 5
    Recording SVB Macros (Programs): Macros, Master (Log) Macros, and
        Keyboard Macros 6
    STATISTICA Visual Basic Editor and Debugger 7

2. RECORDING MACROS: AUTOMATIC PROGRAMMING 9
    Overview: Three Types of SVB Macros 11
    1. Analysis Macros 13
        Datafile Selections and Operations 14
        Case Selection Conditions, Case Weights 14
        Handling Output; Sending Results to Workbooks, Reports, etc. 15
        Example 15
    2. Master Macros: Log of an Entire Analysis 17
        Datafile Selections 18
        Data Editing Operations 19
        Recording Consecutive and Simultaneous Analyses 20
        Case Selection Conditions, Case Weights 21
        Handling Output; Sending Results to Workbooks, Reports, etc. 21
        Master Macro Recording and Analysis Macro Recording 21
        Applications 22
    3. Keyboard Macros 22
    Example: Recording an Analysis Macro 22

SVB PRIMER: CONTENTS

3. PROGRAMMING ENVIRONMENT 27
A Simple Message Box, and If..Then..End If Block 29
Basic Rules for Simple SVB Programs 30
Performing Computations, Data Types, Subroutines, Functions 30
Collections vs. Arrays 32
The Variant Data Type 32
Global Variables, Passing Arguments By Value (ByVal) or By Reference (ByRef) 32
Objects, Methods, and Properties; Running *STATISTICA* from inside Excel 34
Calling Functions in External DLLs 36

4. SVB OBJECT MODEL: EXAMPLES 37
Example: A Simple SVB Program to Compute Descriptive Statistics 39
Example: Retrieving a Collection of Spreadsheets 42
Example: Retrieving Output Documents from AnalysisOutput Objects 43

5. LIBRARIES AND MODULES 45
Overview 47
STATISTICA Visual Basic Reference Libraries and Modules 48
Accessing *STATISTICA* Visual Basic Libraries 50

6. COMMON ELEMENTS OF RECORDED SVB PROGRAMS 53
Recorded Macro Programs 55
Debugging a Macro Program 63

7. CUSTOMIZING *STATISTICA* WITH SVB 65
Creating Dialogs in *STATISTICA* Visual Basic 67
A Simple Dialog in *STATISTICA* Visual Basic 67
Servicing Complex Dialogs via Dialog Functions 71
Servicing Option Buttons, List Boxes, Etc. 72
Controlling *STATISTICA* Events with SVB Programs 76
What Are Events? 76
Types of Events 76
Example: Responding to Document-Level Events 77
Example: Responding to Application-Level Events 78
Supported Events 78
Customizing Toolbars and Menus via *STATISTICA* Visual Basic 79

STATISTICA Visual Basic Primer - iii

8. MATRIX AND STATISTICAL FUNCTION LIBRARIES 81
  Include File: STB.svx 83
  A Simple Example: Inverting a Matrix 84

9. INTRODUCTORY EXAMPLES 87
  Displaying a Simple Message Box 89
  Making a Spreadsheet and Filling It with Random Numbers 90
  Displaying a Progress Bar 90
  Making a Histogram with Normal Fit 91
  Placing Results in Workbooks, Reports, Etc., via the RouteOutput Method 92
  Sending Results to a Report Window 93

10. ADVANCED EXAMPLES 95
  Creating and Customizing Graph Objects 98
  Creating a Cell-Function Spreadsheet 104
  SVB Program for a By-Group Analysis 107

APPENDIX
  A. Document-Level Event Commands 125
  B. Application-Level Event Commands 133

CHAPTER 1

STATISTICA VISUAL BASIC (SVB) – OVERVIEW

Creating *STATISTICA* Visual Basic Programs 3
Executing *STATISTICA* Visual Basic Programs 4
Applications for *STATISTICA* Visual Basic
  Programs 5
Structure of *STATISTICA* Visual Basic 5
Recording SVB Macros (Programs) 6
STATISTICA Visual Basic Editor and Debugger 7

CHAPTER

STATISTICA VISUAL BASIC (SVB) – OVERVIEW

The industry standard *STATISTICA* Visual Basic language (integrated into *STATISTICA*) offers incomparably more than just a "supplementary application programming language" that can be used to write custom extensions. *STATISTICA* Visual Basic (SVB) takes full advantage of the object model architecture of *STATISTICA* and allows you to access programmatically every aspect and virtually every detail of the functionality of the program. Even the most complex analyses and graphs can be recorded into Visual Basic (SVB) macro programs and later be run repeatedly or edited and used as building blocks of other applications. *STATISTICA* Visual Basic adds an arsenal of more than 10,000 new functions to the standard comprehensive syntax of Microsoft Visual Basic thus comprising one of the largest and richest development environments available.

Creating *STATISTICA* Visual Basic Programs

There are several methods in which *STATISTICA* Visual Basic programs can be created.

- *Recording a macro.* When you run an analytic procedure (from the **Statistics** menu) or create a graph (from the **Graphs** menu) the Visual Basic code corresponding to all design specifications as well as output options are recorded in the background. Entire interactive *STATISTICA* analysis sessions involving different types of analyses can be recorded via the Master Macro facility [via **Start Recording Log of Analyses (Master Macro)** on the **Tools - Macro** menu]. That code can later be executed repeatedly or edited by changing options, variables, datafiles, optionally adding a user interface, etc.

- *SVB development environment.* Programs can be written from scratch using the *STATISTICA* Visual Basic professional development environment featuring a convenient

CHAPTER 1: SVB OVERVIEW

program editor with a powerful debugger (with breakpoints, etc.), intuitive dialog painter, and many facilities that aid in efficient code building.

- *Visual Basic from other applications.* SVB programs can also be developed by enhancing Visual Basic programs created in other applications (e.g., Microsoft Excel), by calling *STATISTICA* functions and procedures.

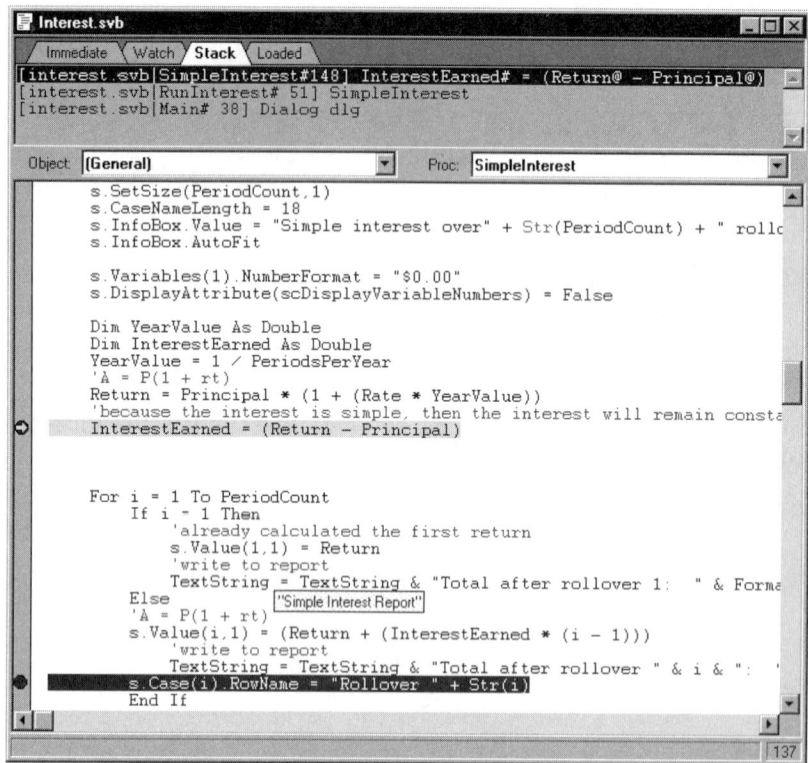

Executing *STATISTICA* Visual Basic Programs

STATISTICA Visual Basic (SVB) programs can be run from within *STATISTICA*; because of the industry standard compatibility of *STATISTICA* Visual Basic and the various libraries of the *STATISTICA* system (accessible to Visual Basic), you can also access *STATISTICA* Visual Basic functions from any other compatible environment (e.g., Microsoft Excel, Word, or a stand-alone Visual Basic language). The large library of *STATISTICA* functions (more than 10,000 procedures) is transparently accessible not only to Visual Basic (either the one that is

4 – *STATISTICA* Visual Basic Primer

built in or a different one), but also to calls from any other compatible programming language or environment such as C/C++, Java, or Delphi.

Applications for *STATISTICA* Visual Basic Programs

STATISTICA VB programs can be used for a wide variety of applications from simple macro (SVB) programs recorded to automate a specific (repeatedly used) sequence of tasks, to elaborate custom analytic systems combining the power of optimized procedures of *STATISTICA* with custom developed extensions featuring their own user interface. Scripts for analyses developed this way can be integrated into larger computing environments or executed from within proprietary corporate software systems or intranet or Internet portals. *STATISTICA* VB programs can also be attached to virtually all important "events" in an analysis with *STATISTICA* such as opening or closing files, clicking on cells in spreadsheets, etc.; in this manner, the basic user interface of *STATISTICA* can be highly customized for specific applications (e.g., for data entry operations, etc.).

Structure of *STATISTICA* Visual Basic

STATISTICA Visual Basic consists of two major components: (1) The general Visual Basic programming environment with facilities and extensions for designing user interfaces (dialogs) and file handling, and (2) the *STATISTICA* libraries with thousands of functions that provide access to practically all functionality of *STATISTICA*.

The Visual Basic programming environment follows the industry standard syntax conventions of the Microsoft Visual Basic language; the few differences pertain mostly to the manner in which dialogs are created (see Chapter 7 - *Customizing STATISTICA with SVB*, page 67), and they are designed to offer programmers and developers more flexibility in the way user interfaces are handled in complex programs. In the *STATISTICA* Visual Basic programming environment, dialogs can be entirely handled inside separate subroutines, which can be flexibly combined into larger multiple-dialog programs; Microsoft Visual Basic is form based, where the forms or dialogs, and all events that occur on the dialogs, are handled in separate program units.

Recording SVB Macros (Programs): Macros, Master (Log) Macros, and Keyboard Macros

STATISTICA provides a comprehensive selection of facilities for recording macros (SVB programs) to automate repetitive work or to be used as a means to automatically generate programs for further editing and modification. The macro (*STATISTICA* Visual Basic) programs recorded by these facilities can be saved to be run "as is," or they can be used as the "building blocks" for more complex and highly customized Visual Basic application programs.

There are three general categories of macros that can be automatically created as you run the program:

- Analysis Macros,
- Master Macros (logs of multiple analyses), and
- Keyboard Macros.

All three follow the identical syntax and can later be modified, but because of the different ways in which each of them is created, they offer distinctive advantages and disadvantages for specific applications. See also Chapter 2 - *Recording Macros: Automatic Programming* (page 11) for additional details.

1. Analysis Macros. First, you can record simple Analysis Macros from an analysis to record the settings, selections, and chosen options for that specific analysis. (Note that the term "analysis" in *STATISTICA* denotes one task selected either from the **Statistics** or **Graphs** menu, which can be very small and simple (e.g., one scatterplot requested from the **Graphs** menu), or very elaborate (e.g., a complex structural equation modeling analysis selected by choosing that command from the **Statistics** menu, and involving hundreds of output documents). After selecting any of the statistical commands from the **Statistics** menu or graphics commands from the **Graphs** menu, all actions such as variable selections, option settings, etc. are recorded "behind the scenes;" at any time you can then transfer this recording (i.e., the Visual Basic code for that macro) to the Visual Basic Editor window. Note that the **Create Macro** command is available from every analysis via the **Options** menu or the shortcut menu (accessed by right-clicking the analysis button) when the respective analysis is minimized.

2. Master Macros (Logs). Second, you can record a Master Macro or Master Log of your entire session that can consist of several or many analyses; this recording will "connect" analyses performed with various analysis options from the **Statistics** or **Graphs** menu.

However, unlike simple Analysis Macros, you can turn the recording of Master Macros on or off. The Master Macro recording will begin when you turn on the recording, and it will end when you stop the recording. In between these actions, all file selections and most data management operations are recorded, as are the analyses and selections for the analyses, in the sequence in which they were chosen.

3. Keyboard macros. When you select **Start Recording Keyboard Macro** from the **Tools - Macro** menu, *STATISTICA* will record the actual keystrokes you enter via the keyboard. When you stop the recording, a *STATISTICA* Visual Basic editor window will open with typically a very simple program containing a single `SendKeys` command with symbols that represent all the different keystrokes you performed during the recording session. Note that this type of macro is very simple in the sense that it will not record any context in which the recorded keystrokes are pressed and will not record their meaning (i.e., commands that these keystrokes will trigger), but this feature makes them particularly useful for some specific applications.

STATISTICA Visual Basic Editor and Debugger

The *STATISTICA* Visual Basic environment includes a flexible program editor and powerful debugging tools. These facilities are described in detail in *Editing and Customizing Recorded Macro (SVB) Programs* in the *Electronic Manual*.

When editing macro programs by typing in general Visual Basic commands or program commands specific to *STATISTICA* Visual Basic, the editor displays *type-ahead* help to illustrate the appropriate syntax.

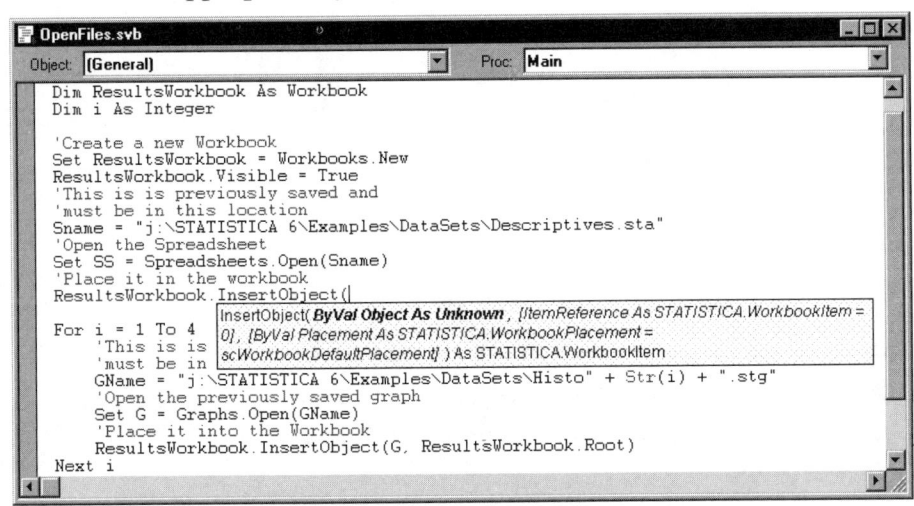

Help on the members and functions for each class (object) is also provided in-line (see Chapter 4 - *SVB Object Model: Examples*, page 39).

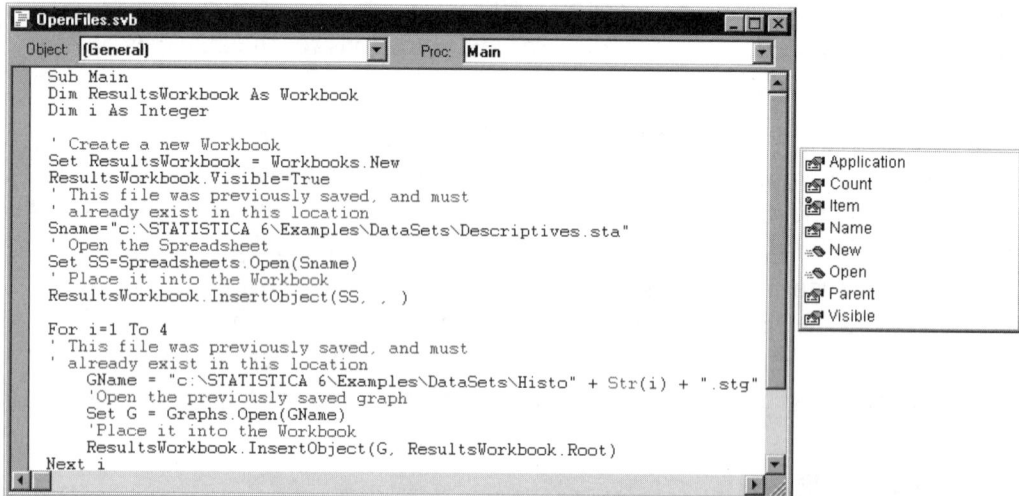

When executing a program, you can set breakpoints in the program, step through line by line, and observe and change the values of variables in the macro program as it is running.

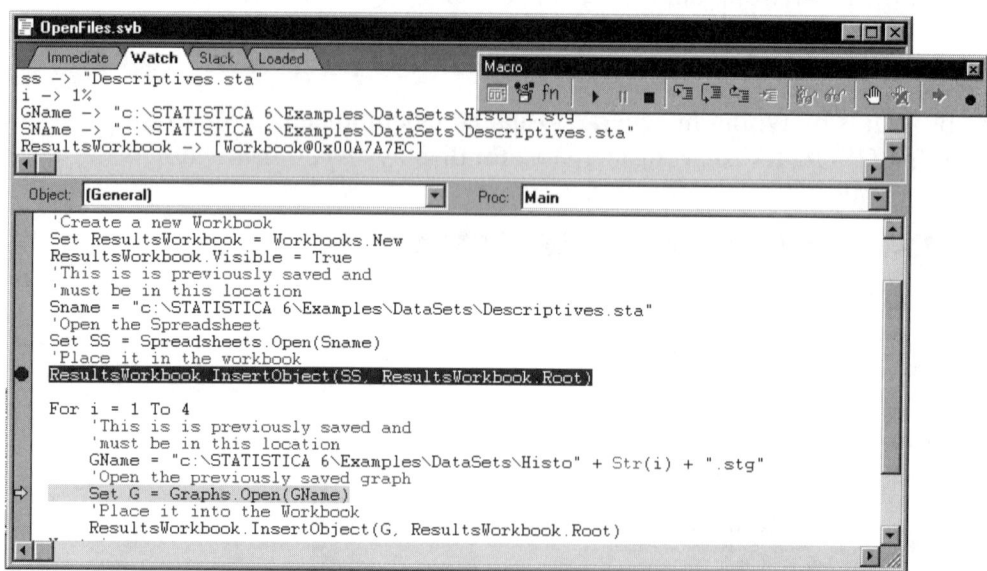

To summarize, *STATISTICA* Visual Basic is not only a powerful programming language, but it represents a very powerful, professional programming environment for developing simple macros as well as complex custom applications.

RECORDING MACROS: AUTOMATIC PROGRAMMING

CHAPTER 2

Overview: Three Types of SVB Macros 11
  1. Analysis Macros 13
  2. Master Macros: Log of an Entire Analysis 17
  3. Keyboard Macros 22
Example: Recording an Analysis Macro 22

CHAPTER 2

RECORDING MACROS: AUTOMATIC PROGRAMMING

STATISTICA provides a comprehensive selection of facilities for recording macros to automate repetitive work or to be used as a means to automatically generate programs for further editing and modification. The macro (*STATISTICA* Visual Basic) programs recorded by these facilities can be saved to be run "as is," or they can be used as the "building blocks" for more complex and highly customized Visual Basic application programs.

Overview: Three Types of SVB Macros

There are three categories of macros that can be automatically created as you run *STATISTICA*:

1. Analysis Macros,
2. Master Macros (logs of multiple analyses), and
3. Keyboard Macros.

All three follow the identical syntax and can be modified at a later time, but because of the different ways in which each of them is created, they offer distinctive advantages and disadvantages for specific applications.

1. Analysis Macros. First, you can record simple Analysis Macros from an analysis to record the settings, selections, and chosen options for that specific analysis. [Note that the term "analysis" in *STATISTICA* denotes one task selected either from the **Statistics** or **Graphs** menu, which can be very small and simple (e.g., one scatterplot requested from the **Graphs** menu), or very elaborate (e.g., a complex structural equation modeling analysis selected by choosing that option from the **Statistics** menu and involving hundreds of output documents).] After selecting any of the statistical options from the **Statistics** menu or graphics options from the **Graphs** menu, all actions such as variable selections, option settings, etc. are recorded "behind the scenes;" at any time you can then transfer this recording (i.e., the Visual Basic code for that macro) to the Visual Basic program editor window. Note that the **Create Macro** command is available from every analysis via the **Options** menu or the shortcut menu (accessed by right-clicking the analysis button) when the respective analysis is minimized.

Also, your choice of datafiles, as well as case selection conditions and the weight variable, are recorded as long as those options are selected in the analysis dialog (not from the **File** menu or the status bar). One such "stand alone" macro is created for each analysis and these macros are not "put together" automatically by *STATISTICA* (in fact they cannot be "mechanically" combined without some editing since each of them represents a stand-alone program that starts with the appropriate declarations, etc.).

2. Master Macros (Logs). Second, you can record a Master Macro or Master Log of your entire session, that may consist of several or many analyses; this recording will "connect" analyses performed with various analysis options from the **Statistics** or **Graphs** menu. However, unlike simple Analysis Macros, you can turn the recording of Master Macros on or off. The Master Macro recording will begin when you turn on the recording, and it will end when you stop the recording. In between these actions, all file selections and most data management operations are recorded, as are the analyses and selections for the analyses, in the sequence in which they were chosen.

The most common application of the Master Macro would be to start *STATISTICA*, start the Master Macro recording by selecting **Start Recording Log of Analyses (Master Macro)** from the **Tools - Macro** menu, and then continue with the analyses. For example, you can compute descriptive statistics, perform some multiple regression analyses, make several histograms and scatterplots, etc. Note that during the analysis, you will see the floating **Record** toolbar to remind you that you are currently recording a Master Macro.

Finally, you stop the recording by either clicking the stop button ■ on the floating **Record** toolbar (see above) or by selecting **Stop Recording** from the **Tools - Macro** menu. At that point, the Visual Basic program that represents all actions performed or selections made during the Master Macro recording will be transferred into a *STATISTICA* Visual Basic editor window. When you run this macro "as is," the exact same analyses will be repeated (with some exceptions resulting from the logic of creating reusable programs from sequences of interactive operations performed by the user, described below).

3. Keyboard Macros. When you select **Start Recording Keyboard Macro** from the **Tools - Macro** menu, STATISTICA will record the actual keystrokes you are entering via the keyboard. The floating **Record** toolbar will be visible as a reminder that a recording "session" is in progress.

When you stop the recording by either clicking the stop button ■ on the floating **Record** toolbar (see above) or by selecting **Stop Recording** from the **Tools - Macro** menu, a STATISTICA Visual Basic editor window will open with typically a very simple program containing a single **SendKeys** command with symbols that represent all the different keystrokes you performed during the recording session.

Note that this type of macro is very simple in the sense that it will not record any context in which the recorded keystrokes are pressed and will not record their meaning (i.e., commands that these keystrokes will trigger), but this feature makes them particularly useful for some specific applications. For example, it can be very useful when it is attached to a keyboard shortcut (select **Customize** from the **Tools** menu to display the **Customize** dialog; then select the **Keyboard** tab, where you can attach the program to a shortcut key) where it could be used, for example, to quickly reenter long text, a formula, a selection of variables, or a large number of options (via keystrokes) on some complex dialog.

1. Analysis Macros

Analysis Macros are the ones that are created automatically and are always being recorded "behind the scenes" whenever you start an analysis from the **Statistics** or **Graphs** menus. Whenever you choose any options from those menus, recording begins; the recording terminates when you exit the analysis, that is, when you click the final **Cancel** button to close the startup dialog. At that point, the recording is discarded and "forgotten" (unlike Master Macros).

There are several things to remember when using the Analysis Macro recording facilities. All of these are consequences of the general "rule": Only actions performed as part of and during the specific analysis being recorded will be reflected in the Analysis Macros.

CHAPTER 2: RECORDING MACROS

Datafile Selections and Operations

The recording of Analysis Macros begins automatically whenever a new analysis is started from the **Statistics** or **Graphs** menu. Anything that "happened" before that is not recorded in the Analysis Macro. Thus, your specific selection of the input datafile is not recorded (the Analysis Macro always assumes that it is to be executed on the current active input datafile) unless that choice is made by clicking the [Open Data] button on the respective analysis dialog. Neither are any operations recorded that you may perform on the input datafile, such as sorting the data, subsetting of variables and/or cases, etc.

Case Selection Conditions, Case Weights

Case selection conditions and case weights are only recorded if they are specified as part of the specific analysis by clicking on the **Select Cases** [SELECT CASES] or **Case Weights** [icon] button of the respective analysis. They will not be recorded in Analysis Macros if they are specified via the respective **Tools** menu options for the input datafile (or if they have been specified prior to the current analysis for the respective input datafile).

Note that in *STATISTICA*, case selection conditions and case weights can be specified either (a) relatively permanently for input datafiles and automatically stored with the files, in which case they will automatically be used by all analyses based on the input datafile, or (b) they can be specified on a per-analysis basis, in which case any new (subsequent) analyses will not use those specifications. In a sense, when you specify case weights or case selection conditions to "belong" to the datafile, those specifications become part of the datafile just like the data (numbers) themselves; when you specify case selection conditions and case weights for a particular analysis, those specifications are only applied to the respective analysis.

To reiterate, case selection conditions and case weights are only recorded in Analysis Macros if they are specified via the respective analysis options (buttons), and regardless of whether or not they are specified for the current analysis only or as a "permanent feature" of the datafile.

Changing current datafile selection conditions and weights. Because case selection conditions and case weights can be recorded into Analysis Macros in the exact same manner as specified via the **Analysis/Graph Case Selection Conditions** and **Analysis/Graph Case Weights** dialogs (when accessed via the respective buttons of the current analysis dialog), it is possible to reset the "permanent" current datafile selection conditions and weights by running

14 – *STATISTICA* Visual Basic Primer

an Analysis Macro. If you run an Analysis Macro that changes the case selection conditions for the current input datafile, then all subsequent analyses will use those specifications, and the results of those analyses will be affected accordingly.

Handling Output; Sending Results to Workbooks, Reports, etc.

Like case selection conditions or case weights, the selections of output options (specifications) on the **Output Manager** are only recorded if they are made from the respective analysis that is being recorded (via the **Output** options on the respective analysis **Options**  menu); changes in the **Output Manager** are not recorded if they are made globally for all analyses via the **Tools - Options** menu. In keeping with the logic of Analysis Macros, only actions that are performed as part of a specific analysis are recorded.

For example, if you recorded an Analysis Macro without making any changes (from that analysis, via the **Options - Output** option available on every dialog of the analysis), then no information about specific settings of output options will be recorded. Consequently, when you run the macro, the output will be directed to the place(s) specified by the current default settings for the output. So, for example, while during the recording of a macro, results spreadsheets might have been directed to individual spreadsheet windows; when running the macro, and while different defaults are in place, the results spreadsheets might be directed to workbooks and report windows.

Example

Following is a typical example when this type of macro recording is extremely useful. Suppose you are performing an exploratory analysis, applying different statistical techniques, and looking at different graphs, etc. At some point, you create a scatterplot that is particularly revealing and interesting by selecting **Scatterplots** from the **Graphs** menu. You would like to "remember" and document exactly how you created that particular graph, so you select **Create Macro** from the shortcut menu (displayed by right-clicking on the analysis button) to record everything that you have done in this particular analysis (e.g., via **Graphs - Scatterplots**).

CHAPTER 2: RECORDING MACROS

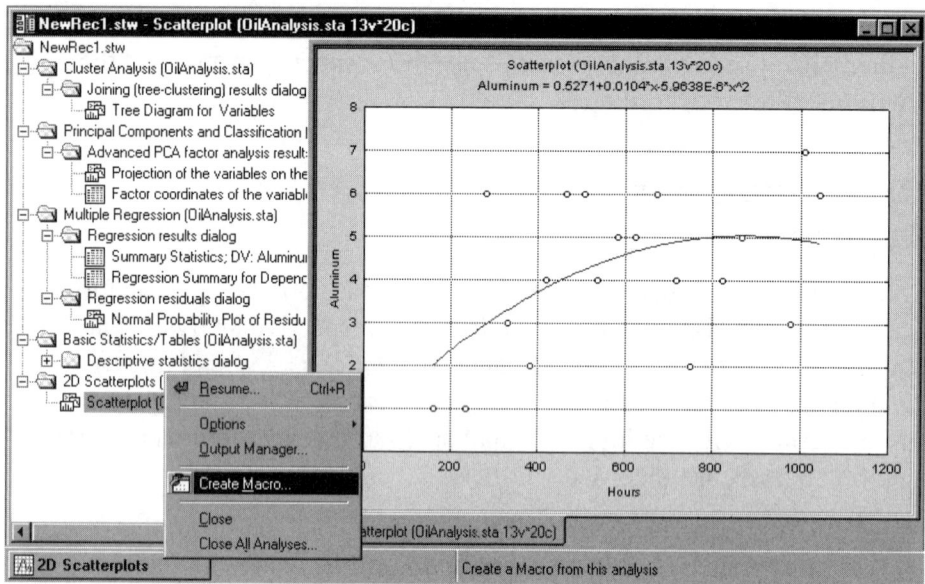

The macro that is recorded will reflect all settings for this particular analysis. In other words, it will not record any multiple regression analyses you may have run or other graphs you have made. It will record only the settings and selections for this particular analysis (**Scatterplots**). The recorded macro may look like this:

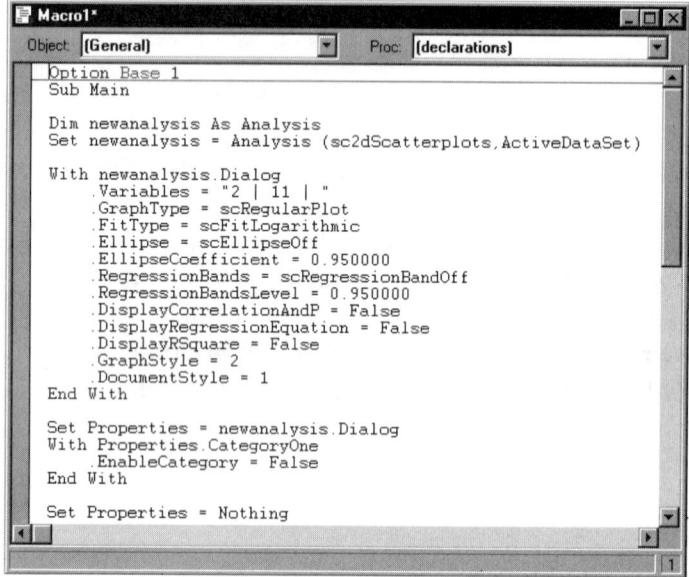

16 – *STATISTICA* Visual Basic Primer

As you can see, the recording started with this particular analysis, by creating a new analysis object for the 2D scatterplot (for details, see also Chapter 3 - *Programming Environment*, page 29, and Chapter 4 - *SVB Object Model: Examples*, page 39). If you run this macro, the exact same graph will be produced; thus, you can save the macro, perhaps with some comments and annotations, to be included in a final report or to be run repeatedly on different datafiles with similar variables.

2. Master Macros: Log of an Entire Analysis

In contrast to simple Analysis Macros, you must explicitly request the recording of Master Macros (logs of entire analysis sessions). To do so, select **Start Recording Log of Analyses (Master Macro)** from the **Tools** - **Macro** menu.

After you start the Master Macro recording, the floating **Record** toolbar will be displayed

to remind you that a *Log* is currently being recorded. From now on, all analyses and most datafile operations are being recorded; that also means that none of the analyses performed prior to starting the recording will become part of the Master Macro. The analyses and chosen results options (e.g., to produce spreadsheets and graphs) are recorded in sequence. Thus, if you are simultaneously running multiple analyses, for example if you computed **Basic Statistics** and at the same time performed some **Multiple Regression** analyses, then the actions taken will be recorded in sequence. As a result, when you play back the Master Macro, you could get first a results spreadsheet computed from **Basic Statistics** followed by a summary table computed in **Multiple Regression** followed by a series of histograms computed via the **Basic Statistics** options, etc.

To stop the recording, click the stop button ■ on the **Record** toolbar; at that point, all recorded Visual Basic syntax (specifying all analyses that were performed during the recording session) will be transferred to a Visual Basic editor window. The Master Macro can then be further edited, saved, or run as is to duplicate the exact sequence of analyses that was performed during recording.

The purpose of the Master Macro is to permit you to create a complete log of all operations. There are a number of issues that you should be aware of when recording and using Master Macros.

Datafile Selections

The recording of the Master Macro begins when you select **Start Recording Log of Analyses (Master Macro)** from the **Tools - Macro** menu. If you select an input datafile after you start the recording, then the selection of that file will become part of the Master Macro. For example, suppose you open a datafile and then start a **Multiple Regression** analysis; the recorded macro will include the following lines:

```
...
Dim S1 As Spreadsheet
Set S1 = Spreadsheets.Open ("c:\datasets\OilAnalysis.sta")
S1.Visible = True
Dim newanalysis2 As Analysis
Set newanalysis2 = Analysis (scMultipleRegression,S1)
...
```

The input datafiles selection was explicitly recorded, and the **Multiple Regression** was initialized using the datafile that was explicitly opened for this analysis. Hence, if you run this Master Macro, the analysis will be performed on the same datafile, i.e., it will be loaded from the disk prior to the **Multiple Regression** analysis (for program details, see also Chapter 3 - *Programming Environment*, page 29, and Chapter 4 - *SVB Object Model: Examples*, page 39).

Now suppose you start the recording of the Master Macro after you have already selected and opened an input datafile. The same analysis might be recorded like this.

```
...
Dim S1 As Spreadsheet
Set S1 = ActiveSpreadsheet
Dim newanalysis2 As Analysis
Set newanalysis2 = Analysis (scMultipleRegression,S1)
...
```

Now the recording started by defining as the input datafile the currently active input spreadsheet. If you run this Master Macro, the **Multiple Regression** analysis will be performed on the currently active input datafile, i.e., possibly a different datafile than that used while recording the Master Macro.

Data Editing Operations

Certain data editing operations on an input dataset are recorded as part of Master Macros. As a general rule, most editing operations that are accomplished via selections on respective dialogs (e.g., the **Sort Options** dialog) are recorded; operations that are performed via simple keyboard actions (e.g., deleting a data value by pressing the DELETE key on your keyboard) or interactive operations performed on the spreadsheet (that typically depend on the current context) are usually not recorded. Also keep in mind that the Master Macro recording only starts when the recording is explicitly requested, so any editing operations that are performed prior to the start of the recording of course will not be reflected in the Master Macro.

Here is a list of data editing operations that are recorded into Master Macros (note that most operations available on the **Data** menu are recorded):

- Sorting
- Creating subsets of cases or variables
- Changes in the variable specifications
- Data transformations via formulas entered into the specifications dialogs of the respective variables
- Adding, moving, copying, and deleting variables
- Adding, moving, copying, and deleting cases
- Ranking of data
- Recoding operations
- Automatic replacement of missing data operations
- Shifting of data
- Standardizing data
- Date operations

Recordable operations performed outside the **Data** menu include:

- Creation of new datafiles (spreadsheets)
- Opening datafiles
- All **Output Manager** options

- Filling data ranges with random values (note that this operation is recorded even though it is accomplished without going through dialogs, but by selecting it from the *Edit* menu)
- Entering case selection conditions on a global and local (one analysis) level
- Entering case weight variables on a global and local (one analysis) level

Note that recording of certain operations as part of the Master Macro logs might lead to creating ambiguous and context-dependent solutions, so they are excluded from the list of recordable tasks. This includes such operations as:

- Editing values in the spreadsheets
- Selections of cells
- Clipboard based copy, paste, and delete operations (note that this does not include the copy, move, and delete cases and variables operations listed above)
- Generally, anything not on the list of recordable data management operations (see above).

Keep these issues in mind when recording a Master Macro with data editing operations that are necessary before subsequent analyses are performed.

Recording Consecutive and Simultaneous Analyses

To reiterate, Master Macro recordings will reproduce the exact sequence of analyses and output choices (of results spreadsheets or graphs) made during the analysis. This facility provides great flexibility and even allows you to "string together" analyses so that the first one computes certain results, while the second one analyzes those results further. For example, you could first perform a *Multiple Regression* analysis on a particular datafile, then use the option *Save* on the results dialog of the *Multiple Regression* analysis to create a stand-alone input spreadsheet of predicted and residual values from the analysis, and then compute *Basic Statistics* on the numbers in that spreadsheet. The Master Macro recording of that sequence of analyses would look something like this (for program details, see also Chapter 3 - *Programming Environment*, page 29, and Chapter 4 - *SVB Object Model: Examples*, page 39):

```
...
Dim S1 As Spreadsheet
Set S1 = Spreadsheets.Open ("c:\STATISTICA\OilAnalysis.sta")
S1.Visible = True
Dim newanalysis1 As Analysis
Set newanalysis1 = Analysis (scMultipleRegression,S1)
...
Dim newanalysis2 As Analysis
Set newanalysis2 = Analysis (scBasicStatistics,ActiveDataSet)
```

```
With newanalysis2.Dialog
    .Statistics = scBasDescriptives
End With
...
```

Note how consecutive analyses (objects) are enumerated as **newanalysis1** and **newanalysis2**. The first one (**Multiple Regression**) is initialized with an explicit input datafile; the second one is initialized with the currently active dataset. When recording complex sequences of analyses like these where results of one analysis serve as the input for subsequent analyses, extra care must be taken to review and, if necessary, edit the final macro before running it to ensure that the intended sequence of **ActiveDataSet**s are chosen by the respective analyses.

Case Selection Conditions, Case Weights

Like Analysis Macros, when case selection conditions or case weights are specified during an analysis, those actions are properly recorded in the Master Macro. However, in addition, when you specify case selection conditions and case weights globally for the datafile (i.e., outside any specific analysis), those actions are recorded as well.

Handling Output; Sending Results to Workbooks, Reports, etc.

Like Analysis Macros, when case output options are changed during an analysis, those selections are recorded in the macro (see also page 41). In addition, if global output defaults are changed via the **Tools - Options** menu (on the **Output Manager** tab of the **Options** dialog), then those choices are recorded into the Master Macro as well. Therefore, if you started a Master Macro, then set the global **Output Manager** option to direct all results spreadsheets and graphs to separate workbooks (for each analysis) as well as to reports, then those selections will be recorded in the Master Macro and reproduced when you execute that macro.

Master Macro Recording and Analysis Macro Recording

The two major modes of macro recording – Master Macros and Analysis Macros, which are always being recorded in the "background" – can be used simultaneously. In other words, you can make Analysis Macros while recording a Master Macro. However, note that the action of creating the Analysis Macro is not itself recorded in the Master Macro.

Applications

Master Macro recording is extremely useful if you want to produce a complete log of all of your analyses. By recording Master Macros that contain specific information about the input datafile, global case selection conditions, etc., you are able to reproduce exactly possibly long "sessions" of analyses you performed interactively. At the same time, through careful planning of when to exactly start the recording, you can also make your long Master Macro of some complex analyses "general," that is, applicable to any datafile that is the current active dataset (see *Datafile Selections* above).

3. Keyboard Macros

When you select **Start Recording Keyboard Macro** from the **Tools - Macro** menu, *STATISTICA* will record the actual keystrokes you enter via the keyboard. When you stop the recording, a *STATISTICA* Visual Basic editor window will open with typically a very simple program containing a single `SendKeys` command with symbols that represent all the different keystrokes you performed during the recording session. Note that this type of macro is very simple in the sense that it will not record any context in which the recorded keystrokes are pressed and will not record their meaning (i.e., commands that these keystrokes will trigger), but this feature makes them particularly useful for some specific applications.

Example: Recording an Analysis Macro

When running analyses in *STATISTICA*, all options and output choices are automatically recorded; when you create an Analysis Macro (via **Options - Create Macro**), the complete recording of all your actions are translated into a *STATISTICA* Visual Basic program that can be run to recreate the specific analysis. Thus, programming in *STATISTICA* can be extremely simple.

Analysis Macros will record the settings, selections, and chosen options for a specific analysis. Note that the term "analysis" in *STATISTICA* denotes one task selected either from the **Statistics** or **Graphs** menu, which can be very small and simple (e.g., one scatterplot requested from the **Graphs** menu), or very elaborate (e.g., a complex structural equation modeling

 CHAPTER 2: RECORDING MACROS

analysis selected by choosing that option from the **Statistics** menu and involving hundreds of output documents. *STATISTICA* also provides facilities to record Keyboard Macros as well as Master Macros (logs). Master Macros will contain logs of your entire analysis; this recording will "connect" analyses performed with various analysis options from the **Statistics** or **Graphs** menus. However, unlike simple Analysis Macros, you can turn the recording of Master Macros on or off. The Master Macro recording will begin when you turn on the recording, and it will end when you stop the recording. In between these actions, all file selections and most data management operations are recorded, as are the analyses and selections for the analyses, in the sequence in which they were chosen.

The following example illustrates how to create an Analysis Macro from a simple analysis:

Start *STATISTICA*, open the example datafile **Exp.sta**, and select **Basic Statistics/Tables** from the **Statistics** menu. On the **Basic Statistics and Tables (Startup Panel)**, select **Descriptive statistics** and click the **OK** button. On the **Descriptive Statistics** dialog, click the **Variables** button and select all variables for the analysis. Then click the **Summary: Descriptive statistics** button to display the descriptive statistics summary spreadsheet. When the spreadsheet is displayed, the **Descriptive Statistics** dialog is minimized on the **Analysis** bar. Click the **Descriptive Statistics** button on the **Analysis** bar to maximize the dialog. Next select the **Normality** tab, and click the **Histograms** button; histograms for all variables will be computed.

The *STATISTICA* Visual Basic code that will reproduce this analysis, which has been "accumulated" by the program, can be displayed in the *STATISTICA* Visual Basic Editor in one of two ways. If the current analysis dialog is minimized, right-click on the button representing the current analysis on the **Analysis** bar, and select **Create Macro** from the resulting shortcut menu.

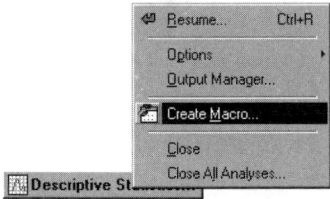

Alternatively, with the analysis dialog maximized, click the [Options] button and select **Create Macro** from the resulting menu.

STATISTICA Visual Basic Primer – 23

CHAPTER 2: RECORDING MACROS

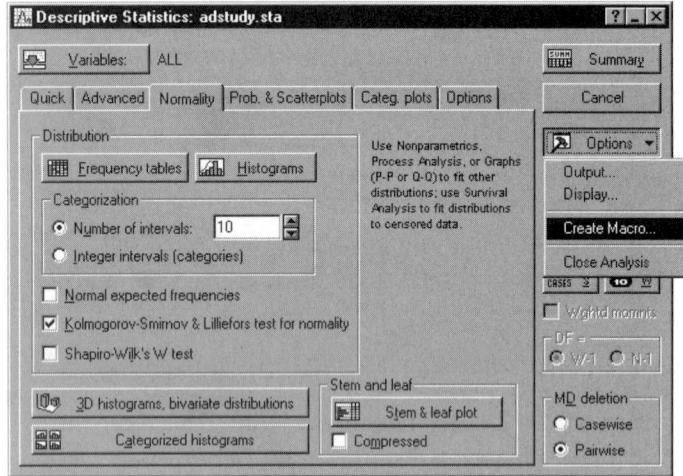

After you select this option, the **New Macro** dialog is displayed. Call the new macro *MyMacro*. Click the **OK** button, and the new macro will be displayed in a stand-alone window. Stand-alone macro programs have to be loaded before they can run, while global macro programs (created by selecting **Save As Global Macro** from the **File** menu) become part of *STATISTICA*; that is, they will automatically be loaded the next time you run *STATISTICA*.

By default, global macros will be available in the **Macro Manager** dialog (accessed by selecting **Macros** from the **Tools** - **Macro** menu). Global macros are the same as standard macros or SVB programs except that they are located in the same directory (by default) where the *STATISTICA* startup program file is located. When you select **Save as Global Macro** from the **File** menu, this directory will automatically be preselected. Save your frequently used macros via **Save as Global Macro** so that they are readily available when you start up *STATISTICA* the next time. Of course, you can always later and "by hand" (e.g., via the Windows operating system functions) move macros in and out of the startup (global macro) directory.

CHAPTER 2: RECORDING MACROS

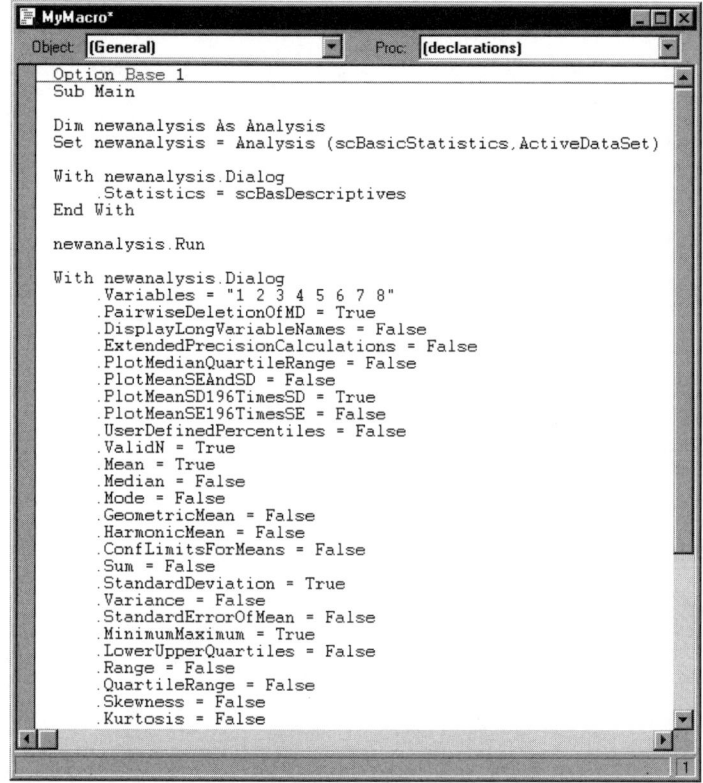

To run this macro, press F5 or click the ▶ **Run Macro** button on the **Macro** toolbar.

To learn more about the details of the syntax of the recorded programs, follow the descriptions below; in particular, see Chapter 6 - *Common Elements of Recorded SVB Programs* (page 55); to learn about Master Macros, which record all analyses and most file operations in an interactive work session, see the discussion earlier in this section (starting on page 9).

PROGRAMMING ENVIRONMENT

A Simple Message Box, and If..Then..End If
  Block 29
Basic Rules for Simple SVB Programs 30
Performing Computations, Data Types,
  Subroutines, Functions 30
Collections vs. Arrays 32
The Variant Data Type 32
Global Variables, Passing Arguments By Value
  (ByVal) or By Reference (ByRef) 32
Objects, Methods, and Properties; Running
  STATISTICA from inside Excel 34
Calling Functions in External DLLs 36

CHAPTER 3

PROGRAMMING ENVIRONMENT

As you will see in the following sections, *STATISTICA* Visual Basic is very similar to Microsoft Visual Basic, as well as the Visual Basic language available in some other advanced (Microsoft Windows) applications (e.g., Microsoft Excel).

A Simple Message Box, and If..Then..End If Block

The following program illustrates various general features of the Visual Basic (VB and SVB) language. This simple example program will display the following dialog and then one of two message boxes depending on the user's action.

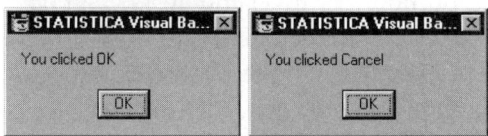

In *STATISTICA*, select **File New**. On the **Create New Document** dialog, select the **Macro (SVB) Program** tab, and create a macro called **Overview 1**.

```
Sub Main
' NOTE: The next statement brings up a message box.
    If MsgBox ("Hello! Click OK or Cancel", _
        vbOkCancel)=vbOK Then
      MsgBox "You clicked OK"
```

STATISTICA Visual Basic Primer – 29

CHAPTER 3: PROGRAMMING ENVIRONMENT

```
        Else
          MsgBox "You clicked Cancel"
        End If
End Sub
```

To run this program, click the ▶ button on the **Macro** toolbar, press F5, or select **Run Macro** from the **Run** menu.

Basic Rules for Simple SVB Programs

- *Main program:* At a minimum, every program consists of the `Main` routine, which is declared as `Sub Main` at the beginning, and terminated with an `End Sub` at the end (*events* can be customized via subroutines with specific names relating to the particular event in question; see also page 75).

- *Comments:* All lines that start with a single quotation mark are interpreted as comments.

- *Continuation lines:* You can break single commands into multiple lines by terminating each line with an underscore ("_"; which must be separated from the preceding text by a blank space).

- *Help on SVB keywords:* At any time, to learn more about the different keywords and statements used in a program, you can highlight the respective text, and then press F1 to display the general SVB help text explaining the syntax for the respective keyword or statement and provide simple examples on how to use them.

Performing Computations, Data Types, Subroutines, Functions

The following program illustrates how variables (for storing numbers or text) and arrays can be declared and used. It also illustrates how subroutines (functions) can be declared and how arguments can be passed to functions. In *STATISTICA*, select **File New**. On the **Create New Document** dialog, select the **Macro (SVB) Program** tab, and create a macro called *Overview 2*.

```
Sub Main
    Dim x (1 To 10) As Double
    Dim Sum As Double, ResText As String
    Dim i As Integer
    For i =1 To 10
      x(i)=i
    Next i
    Sum=ComputeSumOfSqrs ( LBound (x), _
```

30 – *STATISTICA* Visual Basic Primer

CHAPTER 3: PROGRAMMING ENVIRONMENT

```
                            UBound(x), _
                            x)
      ResText="The sum of the square root of values from " + _
              Str(LBound(x)) + _
              " to " + _
              Str(UBound(x)) + _
              " is " + _
              Str(Sum)
      MsgBox ResText
End Sub
Function ComputeSumOfSqrs (iFrom As Integer, _
                          iTo As Integer, _
                          x() As Double) As Double
      Dim i As Integer
      ComputeSumOfSqrs=0
      For i= iFrom To iTo
           ComputeSumOfSqrs=ComputeSumOfSqrs+x(i)^2
      Next i
End Function
```

Numbers. The `Double` data type and the `Integer` and `Long` data types are probably the ones most commonly used in computations. Variables declared as `Double` can hold (store) real numbers, approximately in the range from ±1.7E ± 308 (approximately 15 digits of precision); variables declared as `Integer` can hold (store) integer numbers in the range from -32,768 to 32,767, and `Long` variables can hold (store) integer numbers in the range from -2,147,483,648 to 2,147,483,647.

Strings. Use the `String` data type to operate on character strings of arbitrary length.

Boolean. The `Boolean` data type can hold two values: *True (1)* or *False (0)*.

Arrays. The example program also illustrates how arrays of values are declared and used in Visual Basic. By default, arrays are 0-referenced; this means that an array declared as `Dim x(5)` actually has six elements: The first element can be referenced as `x(0)`, the second as `x(1)`, and the sixth element as `x(5)`. You can also declare arrays with explicit boundaries; for example, you declare an array as `Dim x(1 to 5)`, then it will only have five elements, with the first element referenced as `x(1)`. You can also place at the beginning of the program the statement `Option Base 1`, which will by default declare all arrays as 1-referenced, i.e., with a lower array bound of *1* (this option is inserted by default into recorded macros).

Assigning objects to variables. When assigning objects to variables (see *Objects, Methods, and Properties*, page 34), you need to use the syntax `Set Variable = Object`; for example:

CHAPTER 3: PROGRAMMING ENVIRONMENT

```
Sub Main
    Dim wb As Workbook
    Dim ss As Spreadsheet
    Dim g As Graph
    Set wb=ActiveWorkbook
    Set ss=Spreadsheets.Open( _
      "j:\Statistica\Examples\Datasets\Adstudy.sta")
    ss.Visible=True
    Set g=ActiveGraph
End Sub
```

Collections vs. Arrays

A collection is very similar to an array; however, the collection is an object (see *Objects, Methods, and Properties*, page 34) with various methods that in many instances makes dealing with a collection much more convenient than dealing with an array. In *STATISTICA* Visual Basic, all results spreadsheets and graphs from analyses are returned as collections, which makes the programmatic editing, storing, and further processing of results very convenient.

The Variant Data Type

A variable declared as a `Variant` data type can be empty, numeric, currency, date, string, object, error code, null, or array value. When using SVB to incorporate statistical modules (functions) into a custom program, the `Variant` data type is often useful when, for example, dealing with variable lists, etc. For example, variables can usually be specified as strings (e.g., `.Variables="MyVarName"`), numbers (e.g., `.Variables=2`), or arrays (`.Variables=VarArray`). Note that variables that are not explicitly declared in the program are by default assumed to be of the `Variant` data type.

Global Variables, Passing Arguments By Value (ByVal) or By Reference (ByRef)

ByRef. By default, variables are passed to subroutines and functions by reference. This allows the subroutine or function to update (change) the value of a variable, and to pass the updated value back to the calling routine (from which the respective subroutine or function was called). Thus, if the subroutine or function should be able to change particular values (in the routine from which it was called), pass them by reference (i.e., use the default, or explicitly declare the arguments via the `ByRef` statement in the header of the function). When a variable is passed into a subroutine or function by reference, then the (reference to the) variable itself is passed, and when the value of the variable inside the subroutine changes, then the value in the calling routine (which called the respective routine) changes as well.

```
...
ComputeX x, y1,y2
...
Sub ComputeX(ByRef x As Double, ByVal y1 As Double, _
      ByVal y2 As Double)
' or: Sub ComputeX(x As Double, ByVal y1 As Double, _
      ByVal y2 As Double)
   x=y1+y2
End Sub
```

ByVal. Variables can also be passed to subroutines and functions by value. In a sense, this means, that the value of the variable is passed into the subroutine, and not the variable itself. In practical terms, when the value of a variable that was passed into a subroutine by value is changed, then it will not change in the calling routine (which called the respective subroutine); therefore, arguments passed by value are typically used as input arguments only. For example:

```
...
x = ComputeX(y1,y2)
...
Function ComputeX(ByVal y1 As Double, ByVal y2 As Double) As Double
   ComputeX=y1+y2
End Function
```

Public or global variables. You can also declare variables outside any subroutines and functions, in which case they are "globally" visible to all subroutines and functions in the respective SVB program.

```
Dim x as double, y1 as double, y2 as double

Sub Main
   ComputeX
   ...
End Sub

Sub ComputeX
   x=y1+y2
End Sub
```

Passing arrays. Instead of individual variables, Visual Basic subroutines and functions can also be called with array arguments. Arrays are always passed by reference.

```
Dim xy(3) As Double
...
ComputeX xy
```

```
...
Sub ComputeX (xy() As Double)
  xy(1) =xy(2)+xy(3)
End Sub
```

Objects, Methods, and Properties; Running *STATISTICA* from inside Excel

The exchange of information between different applications is accomplished by exposing those applications to the Visual Basic programs as objects. So, for example, you can run statistical analyses in the *STATISTICA* **Basic Statistics** module from a Visual Basic program in Excel by declaring inside the program an object of type `Statistica.Application`.

Once an object has been created, the Visual Basic program then has access to the properties and methods contained in that object. Properties can be mostly thought of as variables; methods can be mostly thought of as subroutines or functions that perform certain operations or computations inside the respective application object.

Running *STATISTICA* from inside Excel. To illustrate, we will create an example program that can be run from Visual Basic within Excel. After starting Excel, create a new worksheet. From the **Tools** menu, select **Macro - Visual Basic Editor**. From the Visual Basic **Tools** menu, select **References**.

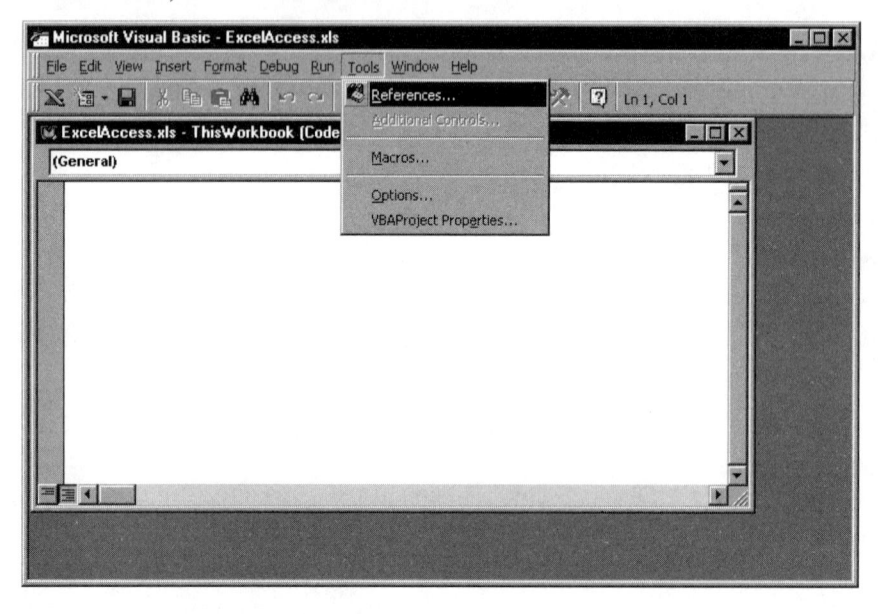

From the **References** dialog, you can select the libraries (objects) that you would like to be visible inside the Visual Basic program. To make *STATISTICA* visible, select the **STATISTICA Object Library** and the **STATISTICA Basic Statistics Library** (for the current version of *STATISTICA*); then click the **OK** button.

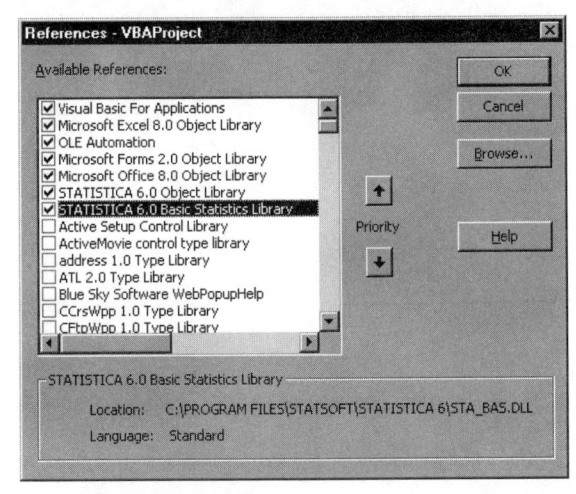

Now type the following program into the program editor.

```
Sub ExcelTest()
Set x = CreateObject("Statistica.Application")
' NOTE: This file may reside in a different directory on
' your installation.
Set a = x.Analysis(scBasicStatistics, _
      "j:\STATISTICA\Examples\Datasets\exp.sta")
   a.Dialog.Statistics = scBasDescriptives
   a.Run
   a.Dialog.Variables = "5-8"
   Set out = a.Dialog.Summary
' Select all rows and columns in the STATISTICA results spreadsheet.
   out.Item(1).SelectAll
   out.Item(1).Copy
   Range("A1").Select
   ActiveSheet.PasteSpecial Format:="Biff4"
End Sub
```

When you run this Visual Basic program from inside Microsoft Excel (Visual Basic Editor), it will paste the results from the **Summary** results spreadsheet of the **Basic Statistics - Descriptive Statistics** analysis into the current Excel spreadsheet.

Calling Functions in External DLLs

You can call from within SVB programs functions defined in external DLLs. Refer to *Calling Functions in External DLLs* in the *Electronic Manual* for calling conventions and the functions declarations required to make external functions visible (useable) in SVB.

CHAPTER 4

SVB OBJECT MODEL: EXAMPLES

Example: A Simple SVB Program to Compute Descriptive Statistics 39

Example: Retrieving a Collection of Spreadsheets 42

Example: Retrieving Output Documents from AnalysisOutput Objects 43

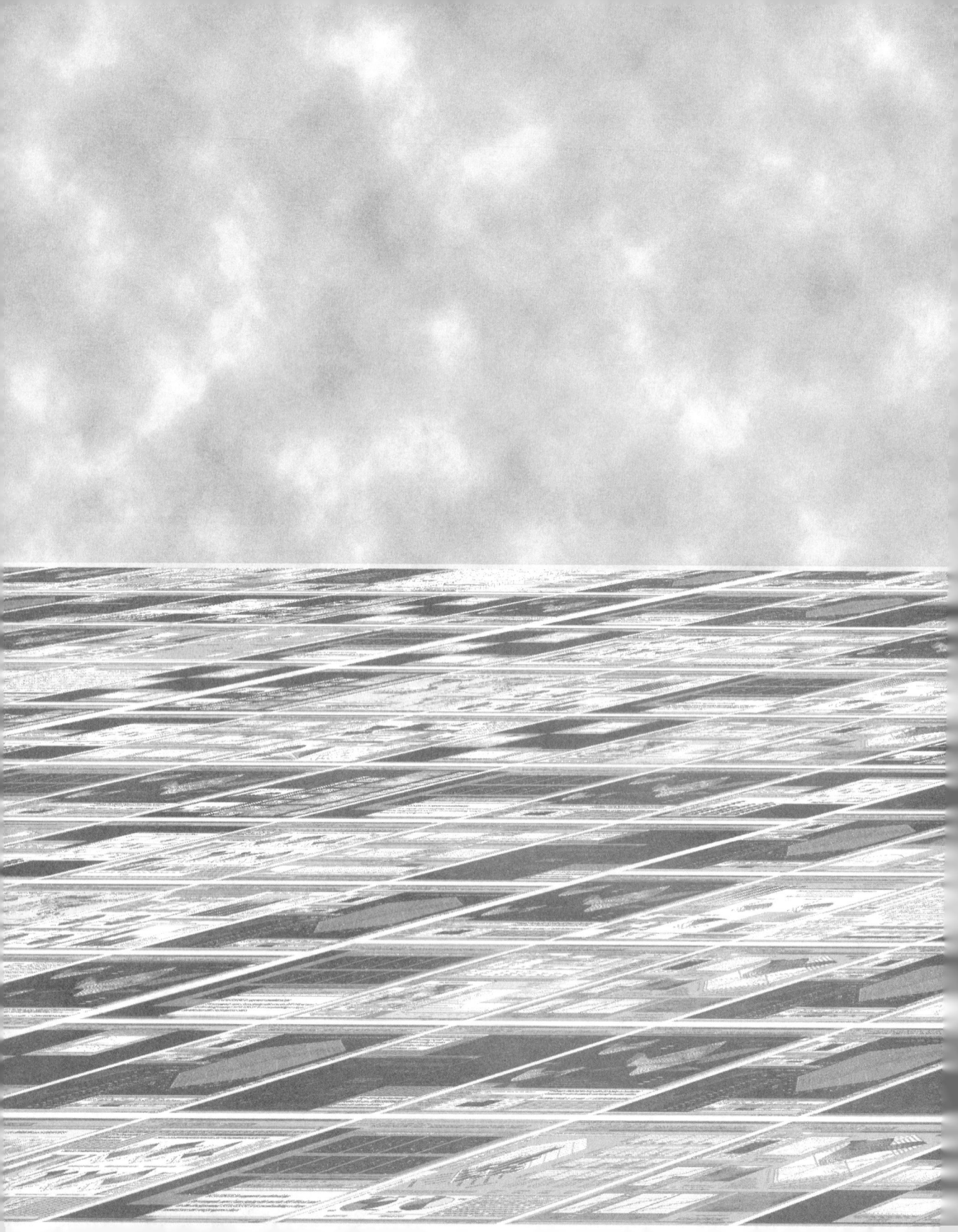

CHAPTER 4

SVB OBJECT MODEL: EXAMPLES

The *STATISTICA* libraries contain all functions that allow you to fully program and customize your *STATISTICA* application; these functions can be called from other applications that support the standard Visual Basic language such as Microsoft Visual Basic, Microsoft Excel, etc.

Example: A Simple SVB Program to Compute Descriptive Statistics

In *STATISTICA*, select **File New**. On the **Create New Document** dialog, select the **Macro (SVB) Program** tab, and create a macro called **Example 1**.

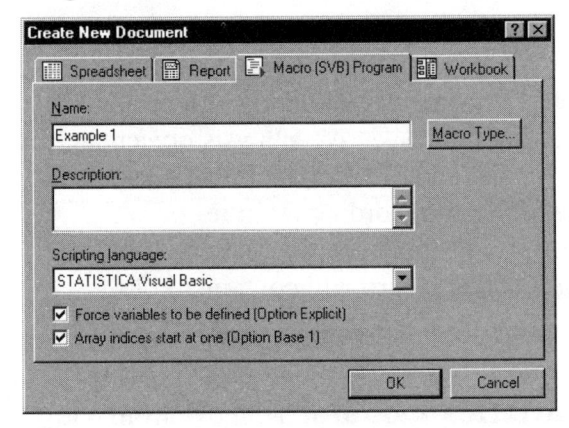

SVB is organized around analysis objects; for example, to run an analysis with the *STATISTICA* **Basic Statistics** module, you would first create an analysis object with the constant `scBasicStatistics` and (optionally) with a datafile name (location of the file

CHAPTER 4: SVB OBJECT MODEL

containing the input spreadsheet). To make access to the thousands of statistical functions and options available in the *STATISTICA* system as convenient as possible, SVB maintains a very close correspondence between the dialogs as they are presented during interactive analyses and the flow of the SVB program. In a sense, once an analysis has been created, such as the analysis via the **Basic Statistics** module in the example, you simply "program the dialogs" for the respective statistical analysis:

```
Sub Main
' NOTE: The datafile may reside in a different directory on
' your installation.  Also, if you recorded this portion of
' the code via a Master Macro [option Tools - Macro - Start
' recording of log of analyses (Master Macro) ], consecutive
' analyses would be enumerated as newanalysis1, newanalysis2,
' etc., and the input datafile spreadsheets would be explicitly
' assigned to variables (objects) S1, S2, etc.
    Set newanalysis = Analysis (scBasicStatistics, _
        "j:\STATISTICA\Examples\Datasets\exp.sta")
    newanalysis.Dialog.Statistics = scBasDescriptives
    newanalysis.Run
    newanalysis.Dialog.Variables = "1-8"
' NOTE: the following line illustrates the simplest way to
' produce a visible results spreadsheet; when recording
' macros from analyses, various additional system settings
' (e.g., output options) are also recorded, and results
' spreadsheets and graphs are typically handled via the
' RouteOutput method.
    newanalysis.Dialog.Summary.Visible = True
End Sub
```

Organization of SVB programs. You can think of each dialog as a property of the (e.g., **Basic Statistics**) analysis, and of each option, selection, etc. on that dialog as a property of that dialog. Thus, you first invoke a module by declaring the respective analysis object, and then set the desired options, etc. as properties of the analysis (and the dialogs of that analysis). Note that when designing actual programs, the automatic macro recording facilities of *STATISTICA* do most of the programming work for you. You simply run the desired analyses interactively, and then create the macro for those analyses; that macro will contain all of the programming code to recreate the analysis step by step, and it can easily be edited (copied, pasted) to create the desired customized application.

Moving between implied dialogs of the statistics module. To move from one dialog to the next when running *STATISTICA* interactively, you click **OK** (to move forward to the next dialog) and **Cancel** (to return to the previous dialog). SVB has two methods to accomplish this that belong to the analysis object: Run (to move forward to the next dialog) and `GoBack` (to return to the previous dialog). Of these, only the `Run` method is illustrated in the simple example.

Creating output documents. Most results from *STATISTICA* analyses are presented in results spreadsheets and graphs. When running *STATISTICA* interactively, you create results spreadsheets and graphs by clicking the respective buttons on the analysis dialogs. In SVB, each results "button" can be (implicitly) clicked by executing the respective method that is part of the respective dialog. For example, practically every analysis dialog in *STATISTICA* has a **Summary** button (identified by the [SUMM] icon) to create the most "important" results from the respective analysis. In SVB, every analysis dialog has a `Summary` method to do the same.

Results spreadsheets and graphs. Each method that can be used to produce results spreadsheets and graphs (e.g., the `Summary` method) will return a collection of spreadsheet objects, graph objects, or spreadsheet and graph objects (see also *Recording Simple Macro (SVB) Programs, Documenting an Analysis* in the *Electronic Manual*). Even if the respective results spreadsheet or graph consists of only a single document, it will still be returned as a collection. Thus, you can use the standard Visual Basic conventions to retrieve individual objects from the collection, etc.

RouteOutput method, and AnalysisOutput objects. When running *STATISTICA* interactively, the output spreadsheets and graphs can be sent to workbooks (by default), stand-alone windows, reports, etc., depending on the selection of options on the **Analysis/Graph Output Manager** dialog. The choices of options in that dialog are implemented by the `AnalysisOutput` object, which can be used as a "container" for the results spreadsheets and graphs. Typically, a summary results spreadsheet or graphs collection would be recorded as:

```
newanalysis.RouteOutput(newanalysis.Dialog.Summary).Visible=True
```

The `RouteOutput` method takes as an argument the `Summary` collection (of spreadsheets, graphs, or both) and places it into the workbook, report, etc. depending on the current selections on the **Analysis/Graph Output Manager** dialog. The `RouteOutput` method actually returns an object of type `AnalysisOutput` which itself has a number of methods and properties to make it fully "programmable."

Note that the actual selections on the **Analysis/Graph Output Manager** dialog are recorded via the `OutputOption` object. Specifically, those options will be recorded either as part of Analysis Macros if they are set or changed via **Options - Output** for the specific analysis, or they will be recorded as part of Master Macros if they are set or changed in a specific analysis as well as via the **Output** tab of the **Options** dialog (accessed from the **Tools – Options** menu). Refer to Chapter 2 - *Recording Macros: Automatic Programming*, page 11, for a more detailed discussion of the macro recording facilities available in *STATISTICA*.

Case selection conditions, weights. When recording a sequence of analyses with different files, case selection conditions, case weights, etc., long sequences of SVB program instructions will be generated to reflect the settings and choices in successive analyses. When you play back such macro recordings, identical analyses will be performed; however, there is an important consideration to remember, or else results produced by a recorded macro can be different from those of the interactive analysis. Case selection conditions and weights that are defined for the input datafile are a property of the input file (document) and not the analysis, and as such, they will not be recorded (a warning message will alert you to this fact whenever you define, for example, case selection conditions for an input datafile), unless:

1. You recorded an Analysis Macro and specified the case selection conditions or case weights from an analysis dialog (via the **Select cases** or **Weights** buttons), i.e., *not* via **Tools - Selection conditions** or **Tools - Weight** (see also page 14 for details);

2. You recorded a Master Macro and specified the case selection conditions or case weights either from an analysis dialog or via the **Tools** menu commands *after* you started the Master Macro recording (via **Tools - Macro - Start Recording Log of Analyses (Master Macro)**; see also page 21 for additional details).

In other words, case selection conditions or case weights defined for a datafile can be thought of as the actual data (numbers) themselves, i.e., they "belong" to the datafile and describe the nature of the data (instead of the nature of a specific analysis). So, for example, when you want to create a recording to document a particular analysis that produced specific results, you may want to explicitly set any case selection conditions or case weights as part of your analyses to ensure that they are properly recorded. Refer to Chapter 2 - *Recording Macros: Automatic Programming*, page 11, for a more detailed discussion of the macro recording facilities available in *STATISTICA*.

Example: Retrieving a Collection of Spreadsheets

In *STATISTICA*, select **File New**. On the **Create New Document** dialog, select the **Macro (SVB) Program** tab, and create a macro. Then type (or paste) in the following program:

```
Sub Main
' NOTE: This file may reside in a different directory on
' your installation.
    Set newanalysis = Analysis (scBasicStatistics, _
        "j:\STATISTICA\Examples\Datasets\exp.sta")
    newanalysis.Dialog.Statistics = scBasFrequencies
    newanalysis.Run
```

```
            newanalysis.Dialog.Variables = "1-8"
            Set s=newanalysis.Dialog.Summary
            s.Visible=False
            MsgBox "Number of Spreadsheets: " + s.Count
            s.Item(s.Count).Visible=True
      End Sub
```

Manipulating and editing STATISTICA documents. Spreadsheets and graphs are only two of the document types that *STATISTICA* uses to handle input and output of statistical analyses. Other document types are (text) reports and workbooks, in which you can organize and manage all other documents. There are a large number of properties and methods available for each of these types of documents that you can use to customize your results or to access aspects of your results for further processing.

Example: Retrieving Output Documents from AnalysisOutput Objects

When running *STATISTICA* interactively, the output spreadsheets and graphs can be sent to workbooks (by default), stand-alone windows, reports, etc., depending on the selection of options on the **Analysis/Graph Output Manager** dialog (accessed by selecting **Output** from the Options button menu). The choices of options in that dialog are implemented by the `AnalysisOutput` object, which can be used as a "container" for the results spreadsheets and graphs. Typically, a summary results spreadsheet or graphs collection would be recorded as:

```
      newanalysis.RouteOutput(newanalysis.Dialog.Summary).Visible=True
```

The `RouteOutput` method takes as an argument the `Summary` collection (of spreadsheets, graphs, or both) and places it into the workbook, report, etc. depending on the current selections on the **Analysis/Graph Output Manager** dialog. The `RouteOutput` method actually returns an object of type `AnalysisOutput` which itself has a number of methods and properties to make it fully "programmable."

Note that the actual selections on the **Analysis/Graph Output Manager** dialog are recorded via the `OutputOption` object. Specifically, those options will be recorded either as part of Analysis Macros, if they are set or changed via **Options - Output** for the specific analysis, or they will be recorded as part of Master Macros, if they are set or changed in a specific analysis as well as via the **Output** tab of the **Tools - Options** dialog. Refer to Chapter 2 - *Recording Macros: Automatic Programming*, page 11, for a more detailed discussion of the macro recording facilities available in *STATISTICA*.

CHAPTER 4: SVB OBJECT MODEL

Suppose in your interactive analysis all results spreadsheets were automatically placed into a workbook; a macro recorded from such an analysis might look as shown in the following example. This example also illustrates how to access the results documents in the AnalysisOutput object (i.e., edit the recorded macro to access the results spreadsheets).

```
Sub Main
' NOTE: This file may reside in a different directory on
' your installation. Also, if you recorded this portion of
' the code via a Master Macro [option Tools - Macro - Start
' Recording of Log of Analyses (Master Macro) ], consecutive
' analyses would be enumerated as newanalysis1, newanalysis2,
' etc., and the input datafile spreadsheets would be explicitly
' assigned to variables (objects) S1, S2, etc.
    Set newanalysis = Analysis (scBasicStatistics, _
            "j:\STATISTICA\Examples\Datasets\exp.sta")
    newanalysis.Dialog.Statistics = scBasFrequencies
    newanalysis.Run
    newanalysis.Dialog.Variables = "1-8"
' Create the Analysis Output object as requested by
' the current settings in the Output Manager.
    Set r=newanalysis.RouteOutput(newanalysis.Dialog.Summary)
' Make sure that the AnalysisOutput object contains
' a Workbook.
    If (r.HasWorkbook=True) then
' We will next find the first results spreadsheet
' (frequency table) that was produced, and extract it
' from the workbook as a stand-alone spreadsheet; note
' that the objects are explicitly dimensioned in the
' following SVB code, to make it more transparent.
        Dim w as Workbook
        Set w=r.Workbook
        Dim wi As WorkbookItem
        Set wi=w.Root.Child
        While (wi.Type<>scWorkbookItemTypeSpreadsheet)
          Set wi=wi.Child
        Wend
        Dim s As Spreadsheet
        Set s=wi.Extract(scWorkbookExtractCopy)
        s.Visible=True
    End If
End Sub
```

CHAPTER 5

LIBRARIES AND MODULES

Overview 47

STATISTICA Visual Basic Reference Libraries
   and Modules 48

Accessing *STATISTICA* Visual Basic Libraries 50

CHAPTER 5

LIBRARIES AND MODULES

Overview

All statistical procedures and graphs and their customization are available as properties or functions to *STATISTICA* Visual Basic (SVB). In order to provide a transparent programming environment for the user, the analysis objects and properties belonging to those objects are arranged to correspond to the respective flow of options and dialogs as if the respective analyses were performed interactively. In addition, a large number of properties, functions, constants, and events (subroutines associated with particular user-initiated events performed on documents, e.g., right-clicking on a results spreadsheet) are available to manage documents, files, and various general analysis options. To review the libraries and scope of the installation of SVB on your machine, you can display the **Object Browser**: Create a new macro (select **New** from the **File** menu; on the **Create New Document** dialog, select the **Macro (SVB) Program** tab, and create a macro), and then select **Object Browser** from the **View** menu.

CHAPTER 5: LIBRARIES AND MODULES

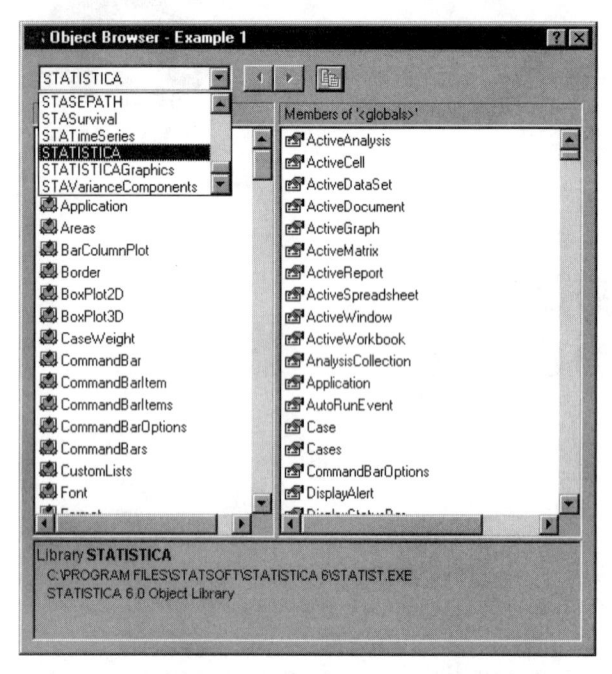

The following is a list of the currently available *STATISTICA* modules and the name of the respective libraries accessible to Visual Basic.

STATISTICA Visual Basic Reference Libraries and Modules

Module (Option)	Library Name	Symbolic Constant
ANOVA*	STAMANOVA	scMANOVA
Basic Statistics	STABasicStatistics	scBasicStatistics
Canonical Analysis	STACanonical	scCanonicalAnalysis
Classification Trees	STAQuickTrees	scClassificationTrees
Cluster Analysis	STACluster	scClusterAnalysis
Correspondence Analysis	STACorrespondence	scCorrespondenceAnalysis
Discriminant Analysis	STADiscriminant	scDiscriminantAnalysis
Distribution Fitting†	STANonparametrics	scDistributions
Experimental Design (DOE)	STAExperimental	scDesignOfExperiments

48 – *STATISTICA* Visual Basic Primer

CHAPTER 5: LIBRARIES AND MODULES

Factor Analysis	STAFactor	scFactorAnalysis
General CHAID Models	STAGCHAID	scGCHAID
General Classification and Regression Trees	STAGTrees	scGTrees
General Discriminant Analysis Models	STAGDA	scGDA
Generalized Additive Models	STAGAM	scGAM
Generalized Linear/Nonlinear Models	STAGLZ	scGLZ
General Linear Models	STAGLM	scGLM
General Partial Least Squares Models	STAPLS	scPLS
General Regression Models	STAGRM	scGSR
Log-Linear Analysis	STALogLinear	scLoglinearAnalysis
Multidimensional Scaling	STAMultidimensional	scMultidimensionalScaling
Multiple Regression	STARegression	scMultipleRegression
Nonlinear Estimation	STANonlinear	scNonlinearEstimation
Nonparametrics	STANonparametrics	scNonparametrics
Principal Components and Classification Analysis[‡]	STAFactor	scAdvancedPCA
Process Analysis Techniques	STAProcessAnalysis	scProcessAnalysis
Quality Control	STAQuality	scQualityControl
Reliability/Item Analysis	STAReliability	scReliabilityandItemAnalysis
Survival Analysis	STASurvival	scSurvivalAnalysis
Time Series	STATimeSeries	scTimeSeries
Variance Components	STAVarianceComponents	scVarianceComponents

*The **ANOVA** results dialog functions are accessible via the **General Linear Models** library.

[†]The **Distribution Fitting** module functions and routines are part of the **Nonparametrics** library.

[‡]The **Principal Components** and **Classification Analysis** techniques are accessible via the **Factor Analysis** library.

CHAPTER 5: LIBRARIES AND MODULES

Note 1: The list of modules and procedures and *STATISTICA* libraries available to Visual Basic is constantly growing. Please check the StatSoft, Inc. website frequently (www.statsoft.com).

Note 2: The *Graphs* menu graphs are all part of the *STATISTICA* main reference library; you can review the respective constants to instantiate a particular analysis (graph type) in the *Object Browser*.

Accessing STATISTICA Visual Basic Libraries

When "launching" a statistical analysis from within a *STATISTICA* Visual Basic program, the first thing to do is to create a new analysis object. If you want to launch a *STATISTICA* analysis from within Visual Basic of another application (e.g., Excel, see page 34), then you first need to load the necessary *STATISTICA* libraries (often via an option called *References* on the *Tools* menu of the application's Visual Basic Editor), and, second, create an object of type `STATISTICA.Application`. You can then create analysis objects that are part of the `STATISTICA.Application` object.

So for example, to access the functions in the *STATISTICA* **Basic Statistics** library from within Microsoft Excel, you would include Visual Basic code like this:

```
Set x = CreateObject("STATISTICA.Application")
' NOTE: This file may reside in a different directory on
' your installation.
Set a = x.Analysis(scBasicStatistics, _
    "j:\STATISTICA\Examples\Datasets\exp.sta")
```

Note that `x` now is an object of type `STATISTICA.Application`; `a` is an object of type `STATISTICA.Application.Analysis`. When you are running a program from within the *STATISTICA* Visual Basic (SVB) program editor, you can omit the explicit declaration of the `STATISTICA.Application` object, and simply write:

```
Set a = Analysis(scBasicStatistics, _
    "j:\STATISTICA\Examples\Datasets\exp.sta")
```

When you run the program from within *STATISTICA*, the program will "know" that the `Analysis` object is part of the current `STATISTICA.Application`. However, you can also create a new `STATISTICA.Application`; for example, try running the following program from within SVB:

```
Sub Main
    Set x = New Application
```

50 – *STATISTICA* Visual Basic Primer

```
        Dim InputDocument As StaDocuments
        Dim InputFile As Spreadsheet
        Set InputDocument=x.Spreadsheets
' NOTE: This file may reside in a different directory on
' your installation.
        Set InputFile=InputDocument.Open( _
            "j:\STATISTICA\Examples\Datasets\exp.sta")
        Set a = x.Analysis(scBasicStatistics, InputFile)
        InputFile.Visible=True
        x.Visible=True
        a.Visible=True
End Sub
```

This program will create (launch) a new *STATISTICA* application; it will then open an input datafile and open the *STATISTICA* **Basic Statistics** module. Finally, all objects declared in this program are made visible, so when it is done, you will see the new *STATISTICA* application, the input datafile (spreadsheet), and the **Startup Panel** for the **Basic Statistics** module.

CHAPTER 6

COMMON ELEMENTS OF RECORDED SVB PROGRAMS

Recorded Macro Programs 55
Debugging a Macro Program 63

CHAPTER 6

COMMON ELEMENTS OF RECORDED SVB PROGRAMS

Recorded Macro Programs

The macro (SVB) programs recorded by *STATISTICA* use several common elements in addition to the properties and methods that are specific to the particular analysis objects.

Creating the analysis object. Each recorded macro begins with declaring the new analysis object:

```
Set newanalysis = Analysis (scBasicStatistics, ActiveDataSet)
```

In this statement, variable `newanalysis` is (implicitly) declared as an `Analysis` object, and the `ActiveDataSet` is a method of the *STATISTICA* application that will retrieve the current active input datafile. Refer to Chapter 5 - *Libraries and Modules*, page 47, for a complete listing of analysis constants and the types of analyses that they create.

If you recorded the macro as a Master Macro, then consecutive analyses (analysis objects) and their datafiles will be "enumerated," so that on play-back identical analyses are performed in the recorded sequence (see also Chapter 2 - *Recording Macros: Automatic Programming*, page 11, for a more detailed discussion of the macro recording facilities available in *STATISTICA*). In that case, the syntax that initiates each analysis may look like this:

```
...
Dim S1 As Spreadsheet
Set S1 = Spreadsheets.Open ("c:\OilAnalysis.sta")
S1.Visible = True
Dim newanalysis1 As Analysis
Set newanalysis1 = Analysis (scMultipleRegression,S1)
...
Dim newanalysis2 As Analysis
Set newanalysis2 = Analysis (scBasicStatistics,S1)
```

...

Note how the first analysis object `newanalysis1` is initiated with an explicit datafile name; the second analysis is initialized with the currently active datafile (`ActiveDataSet`) if, indeed, the datafile selection was not changed during (recorded) interactive analysis.

The With ... End With block; specifying dialogs. This is a shortcut method for setting the options (properties) in a particular dialog. In general, the specification:

```
With newanalysis.Dialog
    .Variables = "5 6 7 8"
    .ConfidenceIntervalForMeansPlot = 95
    .CompressedStemAndLeaf = False
End With
```

is equivalent to specifying these three options like this:

```
newanalysis.Dialog.Variables = "5 6 7 8"
newanalysis.Dialog.ConfidenceIntervalForMeansPlot = 95
newanalysis.Dialog.CompressedStemAndLeaf = False
```

Specifying variables and variable lists. Variables are recorded as a string of variable numbers, which is assigned to the `.Variables` property; when multiple variable lists are recorded (e.g., a list of dependent variables and a list of independent variables), then the string contains the vertical bar character (|) to separate the lists:

```
With newanalysis.Dialog
    .Variables = "5 6 7 8 | 1 2 3 4"
    ...
```

The `.Variables` property can be set by specifying numbers and variable names, as well as variables of type `Variant` (see Chapter 3 - *Programming Environment*, page 29). Like in interactive analyses, in order to select all variables from the current datafile into the respective variable list, you can also use the asterisk "*" character (e.g., to specify a single list containing all variables, you could use `.Variables = "*"`).

Specifying lists of codes. Several types of analyses call for the specification of codes. Here is an example taken from a recording of an analysis via the **Breakdown** option in **Basic Statistics**.

```
With newanalysis.Dialog
    .Variables = "5 6 7 8 | 1 2"
    .Codes = "EXPERMTL-CONTROL | MALE-FEMALE"
End With
```

The `.Codes` property is usually set by assigning to it a string. In this specific example, the codes are identified by 8-character short text labels. They can also be specified as code numbers (i.e., as the actual integer codes that were used in the datafile to identify groups):

```
.Codes = "1 2 | 1 2"
```

Like in interactive analyses, in order to select all codes in the respective variables, you can also use the asterisk "*" character:

.Codes = "* | *"

When the text labels contain special embedded characters such as spaces, you will see double quotation marks to distinguish between the different labels:

```
.Codes = """Experimental group""-""Control group"" |
""MaleParticipants""-""Female participants"""
```

Note that the `.Codes` property can be set by specifying numbers or text labels, as well as variables of type `Variant` (see *Programming Environment*, page 27).

Specifying single codes. Some dialogs contain edit fields for specifying single code values to identify a particular group in the data. For example, the ***T-test for Independent Samples by Groups*** dialog in the ***Basic Statistics*** module contains two such edit fields for specifying the numeric codes or text labels for the grouping variable to identify the groups for the *t*-test.

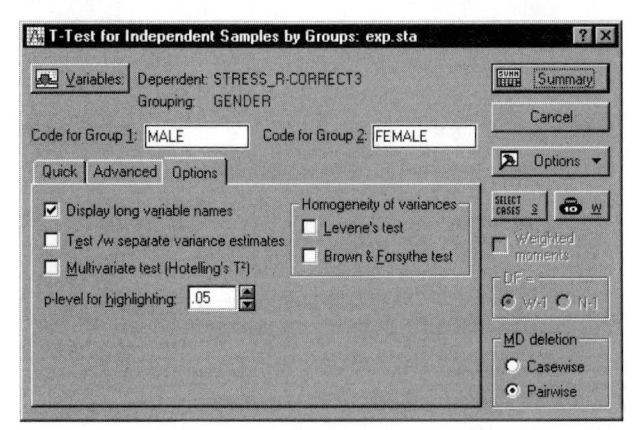

These fields would be recorded in the `With ... End With` block representing this dialog in the following manner:

```
With newanalysis.Dialog
    .Variables = "5 6 7 8 | 2"
    .CodeForGroup1 = "MALE"
```

```
            .CodeForGroup2 = "FEMALE"
             ...
    End With
```

As is the case when specifying lists of codes, the properties `.CodeForGroup1` and `.CodeForGroup2` can also be set by assigning individual values (e.g., `.CodeForGroup1=1`). When the text labels for a variable contain embedded spaces or other special characters, the text values assigned to the respective property will be enclosed in double quotation marks; for example:

```
    .CodeForGroup1 = """MaleParticipants"""
    .CodeForGroup2 = """Female participants"""
```

The single-code selection property can also be set by assignment of a variable of type `Variant` (see *The Variant Data Type*, page 32).

Check boxes, option buttons, numeric input fields. Most dialogs contain various check boxes, groups of option buttons, list boxes, numeric input fields, etc. These options are usually reflected and set inside the respective `With ... End With` block in a straightforward manner. For example, the ***T-test for Independent Samples by Groups*** dialog in Basic Statistics would be recorded as follows:

```
    With newanalysis.Dialog
        .Variables = "5 6 7 8 | 2"
        .CodeForGroup1 = "MALE"
        .CodeForGroup2 = "FEMALE"
        .PairwiseDeletionOfMD = True
        .DisplayLongVariableNames = True
        .TTestWithSeparateVarianceEstimates = False
        .MultivariateTest = False
        .Levene = False
        .BrownForsythe = False
        .PLevelForHighlighting = 0.05
    End With
```

Check boxes and option buttons are checked or cleared by assigning `True` or `False` to the corresponding property. Instead of `True` or `False`, these properties can be set by assigning `1` or `0`, respectively. Numeric input fields are set by assigning the respective values (e.g., `PlevelForHighlighting = 0.05`). As all properties in SVB, properties corresponding to dialog check boxes or option buttons can also be set by assignment of a variable of type `Boolean`, `Integer`, `Long`, or `Variant`, and properties corresponding to numeric input fields can be set by assigning variables of the appropriate type (e.g., `Double` for fields that accept floating point values).

CHAPTER 6: RECORDED SVB PROGRAMS

Combo boxes, list boxes: using symbolic constants. Some dialogs contain combo boxes ("pull-down lists") or list boxes with several items from which to choose. For example, the *Distribution Fitting* module displays the following *Startup Panel*:

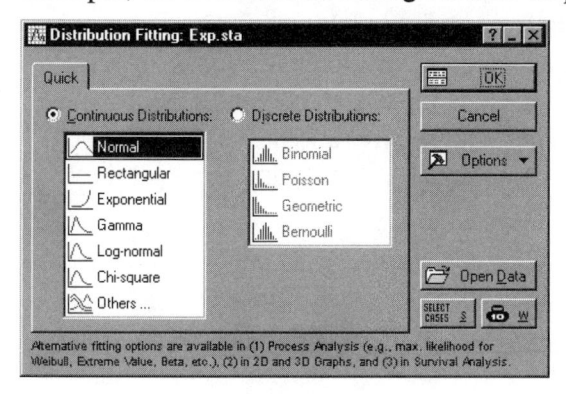

This dialog contains two list boxes, with several distributions from which to choose. If you select the *Normal* distribution, the *Fitting Continuous Distributions* dialog is displayed.

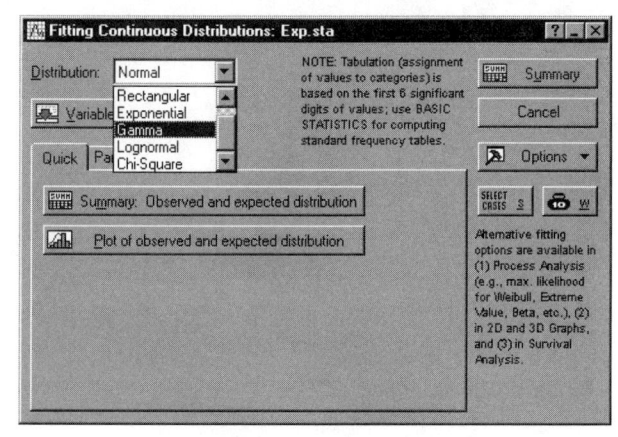

You can still change to a different continuous distribution here by selecting a different option in the *Distribution* list. The choices illustrated above would be recorded as follows:

```
Set newanalysis = Analysis (scDistributions)
With newanalysis.Dialog
    .FitContinuousDistributions = True
    .ContinuousDistribution = scNonNormal
End With
newanalysis.Run
With newanalysis.Dialog
    .Variable = "6"
    .Distribution = scNonGamma
```

STATISTICA Visual Basic Primer – 59

CHAPTER 6: RECORDED SVB PROGRAMS

```
      ...
    End With
```

Note the use of the symbolic constants `scNonNormal` and `scNonGamma`. These are constants that can be substituted by simple values to indicate relative position of the respective choices in the list of options. So for example, `scNonNormal` is the first choice in the list of options in the **Continuous distributions** box on the **Startup Panel**; `scGamma` is the fourth choice in the box on the **Fitting Continuous Distributions** dialog. Because in SVB all enumerations, by default, begin with 0 (and not 1), you could replace these two constants with the values 0 and 3, respectively; so you could also write:

```
    .ContinuousDistribution = 0
    .Distribution = 3
```

The symbolic constants available in a particular dialog can be reviewed in the **Object Browser**.

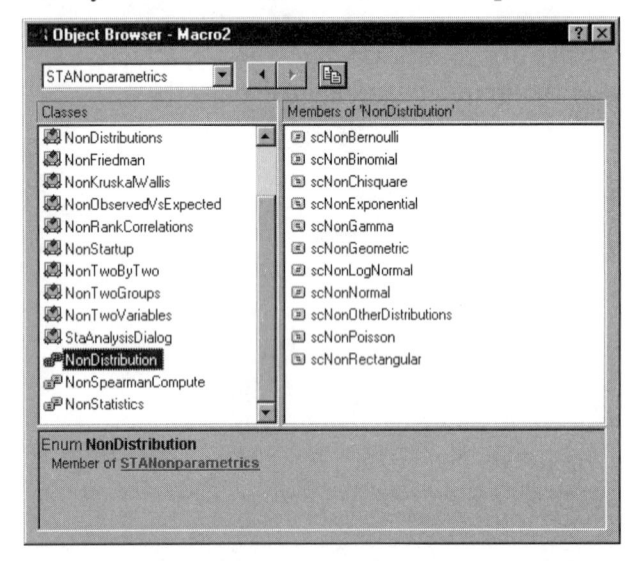

Select the respective object library (note that the **Distribution Fitting** module is contained in the **Nonparametrics** library; see Chapter 5 - *Libraries and Modules*, page 47, for a complete listing of analysis modules and libraries), select an **Enumeration** in the left panel, and then scroll in the right panel to the constants available for that enumeration.

Moving forward (clicking OK); moving backward (clicking Cancel); moving sideways. When running an analysis interactively, a sequence of dialogs is presented with options for each step of the analysis, and you can move back and forth to make modifications as desired. Also, in some cases you can move "sideways" between dialogs; for example, on the **Multiple Regression Results** dialog, there is an option to move to the **Review**

Descriptive Statistics dialog; once you **Cancel** out of that dialog, you will return to **Multiple Regression Results**. Shown below is a part of a typical macro recording those movements, based on a recording of an analysis with the **Multiple Regression** module.

```
Set newanalysis = Analysis (scMultipleRegression)
With newanalysis.Dialog
    .RegressionMode = scRegLinearRegression
    .Variables = "5 | 6 7 8"
End With
newanalysis.Run
With newanalysis.Dialog
    ....
    ....
End With
newanalysis.Dialog.ReviewDescriptiveStatistics
With newanalysis.Dialog
    ....
End With
newanalysis.GoBack
With newanalysis.Dialog
    ...
End With
newanalysis.Run
With newanalysis.Dialog
    ...
End With
newanalysis.GoBack
With newanalysis.Dialog
    ...
End With
```

The `newanalysis.Run` method causes the program to "move forward"; in a sense the `.Run` method "clicks" the **OK** button. The statement `newanalysis.Dialog.ReviewDescriptiveStatistics` is the recording of a "sideways" move; in this particular instance, it moves from the **Multiple Regression Results** dialog to the **Review Descriptive Statistics** dialog. The `newanalysis.GoBack` method is the equivalent of clicking **Cancel**, and thus causes the analysis to "go back" to the previous dialog.

Results variables, results selections. In some instances, when you select an option to create a results spreadsheet or graph, one or more dialogs are displayed to prompt you to select additional specifications for the selected result. For example, when you make **Interaction plots** from the **Statistics by groups** (**Breakdown** analysis results) dialog, and you have multiple dependent variables and multiple categorical predictors selected for the analysis, then a series of dialogs is displayed to (1) select one or more of the dependent variables, and (2) select the

CHAPTER 6: RECORDED SVB PROGRAMS

order of factors for the plot via a two-list selection dialog titled **Arrangement of factors** (use this dialog to assign the factors to the lower and upper *x*-axis labels in the plot). These selections are recorded via the respective dialog's `.ResultsVariables` and `.ResultsSelection` properties, respectively; for example:

```
    ...
    With newanalysis.Dialog
    ...
    End With
    newanalysis.Dialog.ResultsVariables = "6 7"
    newanalysis.Dialog.ResultsSelection = "1 | 2 "
    newanalysis.Dialog.CategorizedMeansInteractionPlots.Visible = True
```

Here the macro recorded the selection of results variables `6` and `7`, and further `ResultsSelection 1` and `2`; i.e., when you run the program, the first element in the first list (of the two-list selection dialog) will be selected, and then the second element in the second list.

ResultsValues. There are some results spreadsheets and graphs that require you to enter a set of values before the results can be computed. For example, the **Multiple Regression Results** dialog contains options for computing predicted values for a set of predictor variable values. When you click the **Predict dependent variable** button on the **Multiple Regression Results** dialog (on the **Residuals/assumptions/prediction** tab), the following dialog is displayed, in which you specify the independent variable values (in this illustration there are three predictor variables **Correct1**, **Correct2**, and **Correct3**).

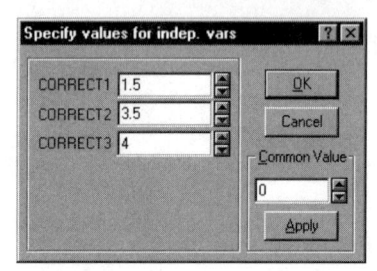

When you click the **OK** button, the predicted value for these independent variable values are computed. When you record a macro to document this analysis, this results option is recorded as follows:

```
    newanalysis.Dialog.ResultsValues = "1.5 3.5 4 "
    newanalysis.Dialog.PredictDependentVar.Visible = True
```

Note that the `.ResultsValues` property can also be set (1) by assigning a single value, in which case all independent variables will be set to that value, (2) by assigning an array of values, or (3) by assigning an array of type `Variant` with values.

Results spreadsheets and graphs; RouteOutput method, and AnalysisOutput objects.

Each button on a results dialog that produces a results spreadsheet or graph (e.g., the **Summary** button) has a corresponding method that is recorded (e.g., the `Summary` method); the method will return a collection of spreadsheet objects, graph objects, or spreadsheet and graph objects (see also *Recording Simple Macro (SVB) Programs, Documenting an Analysis* in the *Electronic Manual*). When running *STATISTICA* interactively, the output spreadsheets and graphs can be sent to workbooks (by default), stand-alone windows, reports, etc., depending on the selection of options on the **Analysis/Graph Output Manager** dialog. The choices of options in that dialog are implemented by the `AnalysisOutput` object, which can be used as a "container" for the results spreadsheets and graphs. Typically, a summary results spreadsheet or graphs collection would be recorded as:

```
newanalysis.RouteOutput(newanalysis.Dialog.Summary).Visible=True
```

The `RouteOutput` method takes as an argument the `Summary` collection (of spreadsheets, graphs, or both) and places it into the workbook, report, etc. depending on the current selections on the **Analysis/Graph Output Manager** dialog. The `RouteOutput` method actually returns an object of type `AnalysisOutput`, which itself has a number of methods and properties to make it fully "programmable."

Note that the actual selections on the **Analysis/Graph Output Manager** dialog are recorded via the `OutputOption` object. Specifically, those options will be recorded either as part of Analysis Macros if they are set or changed for the specific analysis via the **Analysis/Graph Output Manager** dialog (accessed by clicking the **Options** button and selecting **Output**), or they will be recorded as part of Master Macros if they are set or changed in a specific analysis as well as via the **Output** tab of the **Options** dialog (accessed by selecting **Options** from the **Tools** menu). Refer to Chapter 2 - *Recording Macros: Automatic Programming*, page 11, for a more detailed discussion of the macro recording facilities available in *STATISTICA*.

Debugging a Macro Program

You can set individual breakpoints in the program where you want the program to stop temporarily to allow you to look at the values of variables at that point (set and clear breakpoints by pressing F9 or by clicking the **Toggle breakpoint** toolbar button). You can step through the program by pressing SHIFT+F8 or clicking the **Step over** toolbar button.

CHAPTER 6: RECORDED SVB PROGRAMS

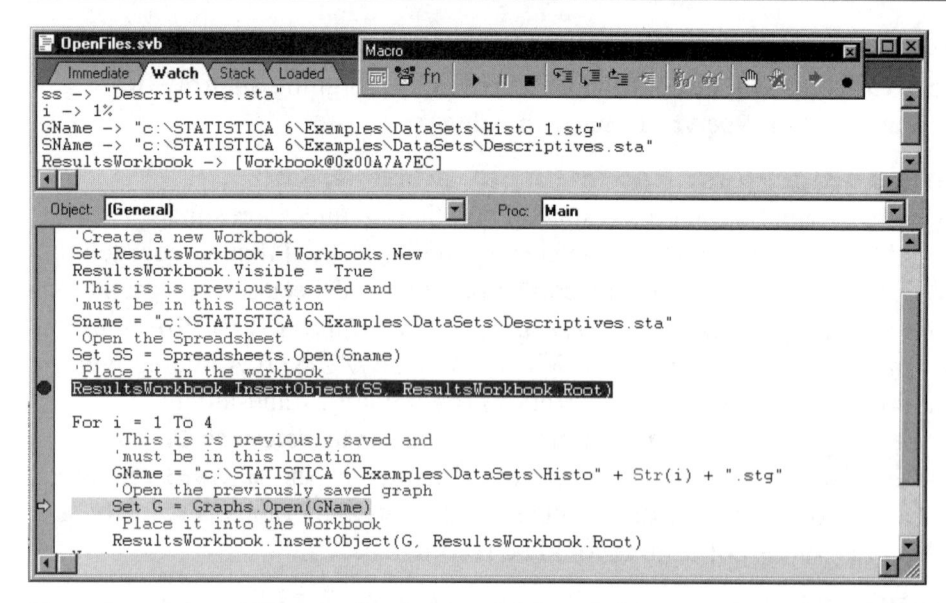

```
OpenFiles.svb
Immediate  Watch  Stack  Loaded
ss -> "Descriptives.sta"
i -> 1%
GName -> "c:\STATISTICA 6\Examples\DataSets\Histo 1.stg"
SNAme -> "c:\STATISTICA 6\Examples\DataSets\Descriptives.sta"
ResultsWorkbook -> [Workbook@0x00A7A7EC]

Object: (General)                    Proc: Main

    'Create a new Workbook
    Set ResultsWorkbook = Workbooks.New
    ResultsWorkbook.Visible = True
    'This is is previously saved and
    'must be in this location
    Sname = "c:\STATISTICA 6\Examples\DataSets\Descriptives.sta"
    'Open the Spreadsheet
    Set SS = Spreadsheets.Open(Sname)
    'Place it in the workbook
    ResultsWorkbook.InsertObject(SS, ResultsWorkbook.Root)

    For i = 1 To 4
        'This is is previously saved and
        'must be in this location
        GName = "c:\STATISTICA 6\Examples\DataSets\Histo" + Str(i) + ".stg"
        'Open the previously saved graph
        Set G = Graphs.Open(GName)
        'Place it into the Workbook
        ResultsWorkbook.InsertObject(G, ResultsWorkbook.Root)
```

Note that the variables in the **Watch** window (tab) not only can be observed, but they can also be changed interactively as the program is running.

CHAPTER 7

CUSTOMIZING STATISTICA WITH SVB

Creating Dialogs in *STATISTICA* Visual Basic 67
Controlling *STATISTICA* Events with SVB
    Programs 76
Customizing Toolbars and Menus via *STATISTICA*
    Visual Basic 79

CHAPTER 7

CUSTOMIZING STATISTICA WITH SVB

The *STATISTICA* Visual Basic environment provides all tools to program complete custom user interfaces. A powerful **User-Dialog Editor** is included to design dialogs using simple operations like dragging buttons to the desired locations. Unlike in Microsoft Visual Basic, the user-defined dialogs are stored along with the program code as data of type **UserDialog**. This method of creating dialogs allows you to create sophisticated user interfaces that can easily be edited in textual form; also, by defining the entire dialog as a variable, you can completely define dialogs inside subroutines, which can be freely moved around the program.

However, user-defined dialogs designed in the *STATISTICA* Visual Basic environment cannot be ported directly to Microsoft Visual Basic, which uses a form-based method of creating dialogs. This is not a serious limitation, though, but rather a "design issue," in the sense that you should decide before embarking on the development of a complex program with extensive custom (user-defined) dialogs which environment you prefer.

Creating Dialogs in *STATISTICA* Visual Basic

A Simple Dialog in *STATISTICA* Visual Basic

The following example illustrates how to create a simple dialog and "service" the user choices on this dialog. Start by creating a new macro: Select **New** from the **File** menu; on the **Create New Document** dialog, select the **Macro (SVB) Program** tab, and create a new macro called *SimpleDialog*.

CHAPTER 7: CUSTOMIZING *STATISTICA* WITH SVB

Creating the dialog. Next select *Dialog Editor* from the *Tools* menu, or click the *Dialog Editor* toolbar button, to display the *UserDialog Editor* dialog.

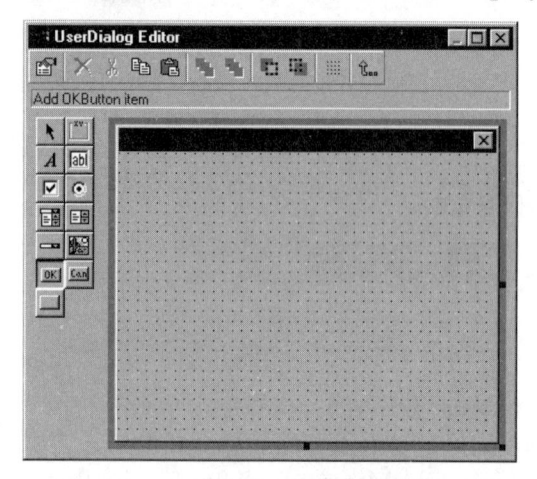

Click the *OK* button on the left side of this dialog, and then click on the upper-right corner of the dialog; an *OK* button will be inserted in that location.

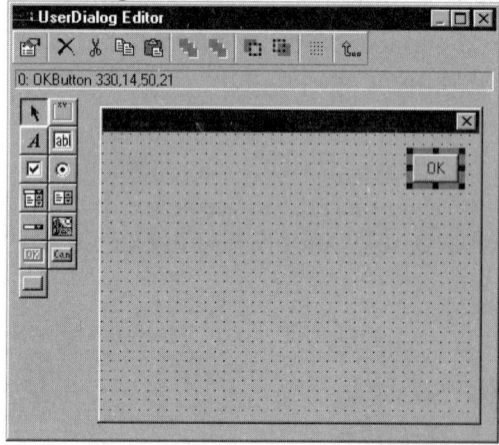

Note that you can further edit the size and location of the *OK* button by clicking it and dragging it to the desired location or resizing it. Repeat these steps for the *Cancel* button and for a user-defined button. To produce the latter, click the blank button toolbar button on the lower-left of the toolbar on the *UserDialog*, and then click the desired location in the dialog editor.

68 – *STATISTICA* Visual Basic Primer

CHAPTER 7: CUSTOMIZING *STATISTICA* WITH SVB

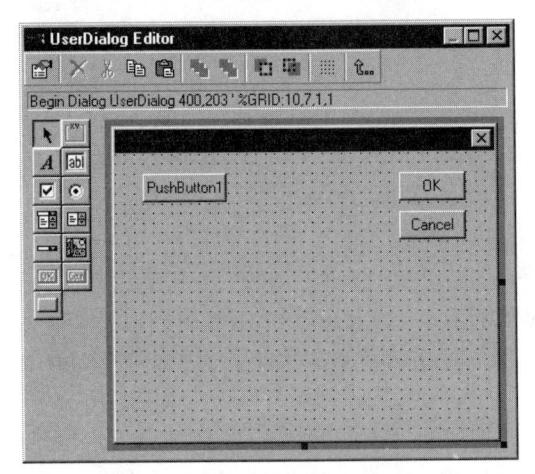

By default, the new button is labeled *Pushbutton1*. To change that, double-click on the button to display the *Edit PushButton Properties* dialog. Then edit the text boxes as follows.

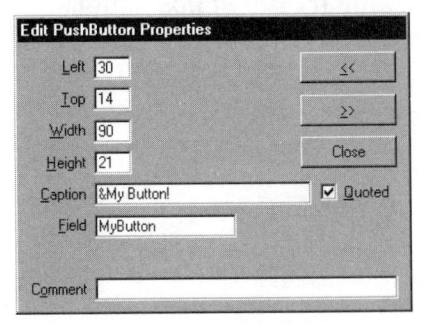

Note that the *Caption* for the new button is *&My Button!* When you click the *Close* button, the actual caption of the button will change to *My Button!*, i.e., with the *M* underlined. Thus, *M* is the keyboard accelerator that will "press" this button. The other important information in the *Edit PushButton Properties* dialog is the "ID" of the button entered into the *Field* box. The ID for this button is *MyButton*; this ID will be referenced throughout the SVB program to service the button, i.e., to identify when the button is clicked and to respond inside the program.

Now close the *Edit PushButton Properties* dialog, and close the *UserDialog Editor* dialog by clicking the toolbar button. The SVB program will now contain the following code.

```
Sub Main
    Begin Dialog UserDialog 400,203 ' %GRID:10,7,1,1
        OKButton 300,14,70,21
        CancelButton 300,42,70,21
        PushButton 30,14,90,21,"&My Button!",.MyButton
    End Dialog
    Dim dlg As UserDialog
```

STATISTICA Visual Basic Primer – 69

CHAPTER 7: CUSTOMIZING *STATISTICA* WITH SVB

```
        Dialog dlg
End Sub
```

Remember that, to learn more about the different keywords and statements used in this program so far, you can always highlight the respective text and then press F1 to display the general SVB help text explaining the syntax for the respective keyword or statement and providing simple examples on how to use them.

Servicing the new dialog; simple dialogs. If you run the program created thus far, the dialog we designed will be displayed on the screen; when you click any button the program will terminate. The next task is to "connect" specific programming instructions to different user actions on the dialog. For example, let us display message boxes to indicate which button the user clicked.

The simplest way to do this is to use the codes returned by the `Dialog` method; this method will return a `0` when **Cancel** is clicked, `-1` for **OK**, and different integers greater than 0 to enumerate the other controls on the dialog. So in this example, clicking the **My Button** button would return a `1`. Here is the program that would service all buttons.

```
Sub Main
      Begin Dialog UserDialog 400,203 ' %GRID:10,7,1,1
         OKButton 300,14,70,21
         CancelButton 300,42,70,21
         PushButton 30,14,90,21,"&My Button!",.MyButton
      End Dialog
      Dim dlg As UserDialog
   Dim ReturnId As Integer
   ReturnId = Dialog (dlg)
   Select Case ReturnId
      Case -1
        MsgBox "The OK button was pressed"
      Case 0       ' Cancel Button
        MsgBox "The CANCEL button was pressed"
      Case 1       ' The 'first' button on the Dialog,
        MsgBox "The MY BUTTON was pressed"
   End Select
End Sub
```

This program uses the standard Visual Basic `Select Case` statement to execute the code for the desired message box, based on the ID number returned by the `Dialog` method; you could of course also accomplish the same effect by using `If ... Then ... Else` statements. (All general Visual Basic statements are documented in the general SVB syntax help in the *Electronic Manual*.)

Servicing Complex Dialogs via Dialog Functions

The simple method of servicing the dialog illustrated in the previous section has the disadvantage that the dialog closes after you click any button. When designing complex user interfaces, you often want the user to set various options on a dialog, and only exit the dialog via the **OK** or **Cancel** button. To accomplish this, you need to specify a dialog function handler subroutine (see also the general SVB help in the *Electronic Manual* for more information on the syntax of dialog functions). Here is the example shown in the previous section, rewritten to use this method of servicing the dialog.

```
Sub Main
Dim ReturnId As Integer
    Begin Dialog UserDialog 400,203, _
        .MyDialogFunction ' %GRID:10,7,1,1
        OKButton 300,14,70,21,.OkButton
        CancelButton 300,42,70,21,.CancelButton
        PushButton 30,14,90,21,"&My Button!",.MyButton
    End Dialog
    Dim dlg As UserDialog
    ReturnId=Dialog(dlg)
End Sub

Private Function MyDialogFunction(DlgItem$, Action%, SuppValue&) As Boolean

    Select Case Action%
    Case 1 ' Dialog box initialization
    Case 2 ' Value changing or button pressed

        MyDialogFunction = True
        Select Case DlgItem

        Case "MyButton"
          MsgBox "The MY BUTTON was pressed"
          MyDialogFunction=True

        Case "OkButton"
          MsgBox "The OK button was pressed"
          MyDialogFunction=True

        Case "CancelButton"
          MsgBox "The CANCEL button was pressed"
```

CHAPTER 7: CUSTOMIZING STATISTICA WITH SVB

```
                MyDialogFunction=False
            End Select
        End Select
End Function
```

When you run this program, you will notice that now you can click the **OK** and **My Button** buttons without closing the dialog. When you click the **Cancel** button, the dialog closes. As noted in the comments of the program, when the `MyDialogFunction` returns `False`, the dialog closes; when it returns `True`, it remains on the screen.

Servicing Option Buttons, List Boxes, Etc.

The example program shown in *Servicing Complex Dialogs via Dialog Functions* (page 71) demonstrates how to service buttons on a custom dialog using a dialog function. The *STATISTICA* Visual Basic general syntax documentation describes how various other dialog controls such as list boxes, etc., can be handled. Here is a fairly elaborate example program to illustrate how to service the various elements commonly used in dialogs.

```
' This program demonstrates how one can service various
' controls in a complex dialog.

Dim ListArray(4) As String
Dim ComboArray(5) As String
Sub Main
    If DisplayDialog = False Then
        GoTo Finish
    End If
    Exit Sub
Finish:
End Sub


Function DisplayDialog As Boolean
    DisplayDialog=True
Begin Dialog UserDialog 390,336,"Demonstating Dialog Controls", _
        .MyDialogFunction ' %GRID:10,7,1,1
            OKButton 300,14,70,21,.OkButton
            CancelButton 300,42,70,21,.CancelButton
            PushButton 30,14,160,21,"&Reset to defaults",.MyButton
            Text 30,49,90,14,"&Combo box:",.TextForMyComboBox
            ComboBox 130,49,160,70,ComboArray(),.MyComboBox
    Text 30,126,100,28,"&Text box (multiple lines):", _
   .TextForTextBox
            TextBox 130,126,160,35,.MyTextBox,1
```

72 – STATISTICA Visual Basic Primer

```
            Text 30,175,90,14,"Value:",.TextForMyTextBox2
            TextBox 130,168,160,21,.MyTextBox2
            Text 30,196,60,14,"List box:",.TextForMyListBox
            DropListBox 130,196,160,91,ListArray(),.MyListBox
            CheckBox 30,231,130,14,"My Checkbox",.MyCheckBox
            GroupBox 30,252,180,63,"&Group Box",.MyGroupBox1
            OptionGroup .MyOptionButtons
                    OptionButton 40,273,150,14,"My Option Button &1"
                    OptionButton 40,294,150,14,"My Option Button &2"
        End Dialog
        Dim dlg As UserDialog
        InitializeUserDialogFields (dlg)
TryAgain:
    DisplayDialog=Dialog(dlg)
    If DisplayDialog=True Then
        If RetrieveUserDialogFields (dlg)=0 Then
            MsgBox "Error in the dialog specs; try again.", vbCritical
            GoTo TryAgain
        End If
    End If
End Function

Private Function MyDialogFunction(DlgItem$, Action%, SuppValue&) As
Boolean

        Select Case Action%

        Case 1 ' Dialog box initialization
        Case 2 ' Value changing or button pressed
                MyDialogFunction = True

                Select Case DlgItem

                Case "MyButton"
                  DlgValue "MyListBox", 0
                  DlgText "MyComboBox", ComboArray(1)
                  DlgText "MyTextBox", _
                        "Initial text for multiple line edit field"
                  DlgText "MyTextBox2",".5"
                  DlgValue "MyCheckBox", True
                  DlgValue "MyOptionButtons", 1
                  MyDialogFunction=True

                Case "MyListBox"
                  MsgBox "New combo box selection:" _
                        + ListArray(SuppValue)
```

```
                    Case "OkButton"
                        MyDialogFunction=False

                    Case "CancelButton"
                        MsgBox "The CANCEL button was pressed"
                        MyDialogFunction=False

                End Select

        End Select

End Function

Sub InitializeUserDialogFields (dlg)
    ListArray(0)="List entry 0"
    ListArray(1)="List entry 1"
    ListArray(2)="List entry 2"
    ListArray(3)="List entry 3"
    ListArray(4)="List entry 4"
    dlg.MyListBox=0

    ComboArray(0)="Combobox entry 0"
    ComboArray(1)="Combobox entry 1"
    ComboArray(2)="Combobox entry 2"
    ComboArray(3)="Combobox entry 3"
    ComboArray(4)="Combobox entry 4"
    ComboArray(5)="Combobox entry 5"
    dlg.MyComboBox=ComboArray(1)

    dlg.MyTextBox="Initial text for multiple line edit field"
    dlg.MyTextBox2=".5"

      dlg.MyCheckBox=True
      dlg.MyOptionButtons=1

End Sub

Function RetrieveUserDialogFields (dlg) As Boolean
On Error GoTo InvalidInput
Dim xval As Double
            RetrieveUserDialogFields=True
        MsgBox "My combo box is set to: " + dlg.MyComboBox
        MsgBox "My multi-line text box is set to: " + dlg.MyTextBox
        xval=CDbl(dlg.MyTextBox2)
        MsgBox "My value was set to: " + Str(xval)
```

```
            MsgBox "My (0-referenced) element chosen in My Listbox is: " _
                + Str(dlg.MyListBox)
            MsgBox "My check box is set to: " + Str(dlg.MyCheckBox)
            MsgBox "My options buttons are set to (0-referenced): " _
                + Str(dlg.MyOptionButtons)
            Exit Function
InvalidInput:
            RetrieveUserDialogFields=False
End Function
```

This example program is fairly complex and illustrates how to interact with the different standard Windows dialog controls. Remember that the general *STATISTICA* Visual Basic help contains detailed documentation for the statements, declarations, and data types used in this program.

Defining dialogs in subroutines; defining sequences of dialogs. In this example program, the dialog is defined in and displayed by a subroutine (function) rather than the main program. Thus, by defining different dialogs in different subroutines, you can build elaborate programs with complex flow control, i.e., sequences of dialogs that depend on prior user choices or results of computations.

Zero referencing in ListBox, DropListBox, and OptionGroup controls. Remember that the elements in `ListBox`, `DropListBox`, and `OptionGroup` controls are zero referenced, i.e., the first element is referenced as number 0 (zero), the second as number 1, etc.

Retrieving numeric values. The program also illustrates how standard text controls can be used to return numeric values (see the `.MyTextBox2` control). Specifically, the program retrieves the user-entered value as text, and later checks whether the text can be converted into a valid value of type `Double`; this is accomplished by defining an `On Error Goto` label in the `RetrieveUserDialogFields` function, where the program control will resume when an error occurs in the `CDbl` conversion function.

Changing controls from inside the dialog function. When you run the program and click the **Reset to defaults** button (with the dialog ID *"MyButton"*), then all fields will be reset to their defaults. This is done via the `DlgText` and `DlgValue` functions, as is illustrated in case *"MyButton"* in the dialog function.

Controlling STATISTICA Events with SVB Programs

What Are Events?

An event is an action that is typically performed by a user, such as clicking a mouse button, pressing a key, changing data, or opening a spreadsheet or workbook. In STATISTICA, certain events are "exposed" to the outside world, i.e., accessible to SVB, and they can be used to customize its behavior. Using programmable events, you can tailor STATISTICA's behavior to your needs. Examples of events applications might include:

- building auditing systems into STATISTICA (by IT departments),
- building interactive demonstration programs based on workbooks,
- building customized user interfaces adhering to specific requirements of a particular application (or a specific company – e.g., to meet specific security requirements).

Events are an important part of the set of tools built into STATISTICA to make it a powerful solution building system.

Types of Events

Document-level events. These events occur for open documents and in some cases for objects within them. For example, the workbook document object can respond to the **Open**, **New**, and **Print** events; the spreadsheet document object can respond to events such as changed data values, double-clicks on a cell, etc. For a listing of the available document-level events, see Appendix A – *Document-Level Event Commands* (page 127).

Application-level events. These events occur at the level of the application itself, for example, when a new spreadsheet, report, graph, or workbook is created. You can create an event handler customizing these actions for all documents of that type. For a listing of the available application-level events, see Appendix B – *Application-Level Event Commands* (page 135).

Example: Responding to Document-Level Events

In this example, we will create an event procedure for events in a *STATISTICA* Spreadsheet; specifically, we will customize a spreadsheet by changing the meaning of the "double-click" on spreadsheet cells. Begin by opening a spreadsheet, e.g., the example datafile **Exp.sta**. Then select **View Code** from the **View - Events** menu.

At this point, the **Document Events** window is displayed allowing you to associate certain events with this spreadsheet. Next, select **Document** in the **Object** box. Then select **BeforeDoubleClick** in the **Proc** box. The **Documents Events** window will now look like this:

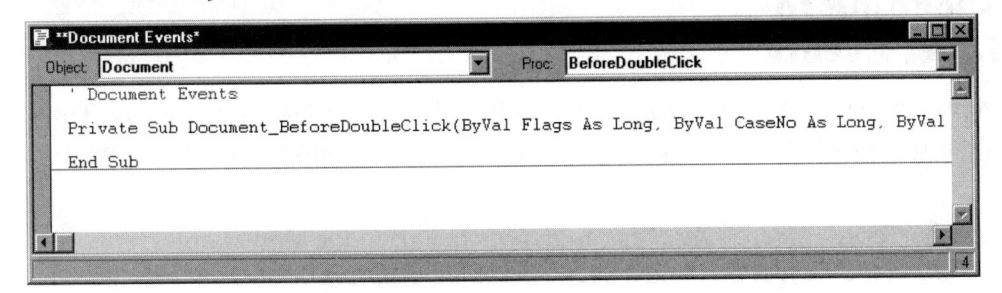

```
**Document Events**
Object: Document                              Proc: BeforeDoubleClick
' Document Events
Private Sub Document_BeforeDoubleClick(ByVal Flags As Long, ByVal CaseNo As Long, ByVal
End Sub
```

The following code will be inserted in this dialog:

```
Private Sub Document_BeforeDoubleClick(ByVal Flags As Long, ByVal
    VarNo As Long, ByVal CaseNo As Long, Cancel As Boolean)
End Sub
```

This code is a placeholder for our event code; change it to the following:

```
Private Sub Document_BeforeDoubleClick(ByVal Flags As Long, _
    ByVal VarNo As Long, ByVal CaseNo As Long, Cancel As Boolean)
    MsgBox "Double-click not allowed here"
    Cancel = True
End Sub
```

Now run this macro to "monitor" the double-click event for the **Exp.sta** spreadsheet. After starting the program, double-click on any cell in the spreadsheet. You will see the following message:

```
STATISTICA Visual Basic
Double-click not allowed here
    [ OK ]
```

CHAPTER 7: CUSTOMIZING *STATISTICA* WITH SVB

As you can see, we have now "protected" the cells of the spreadsheet against the double-click event.

You can save the event handler with the spreadsheet document: Click on the **Document Events** window and select **Save** from the **File** menu. Then close the **Document Events** window, activate the spreadsheet, select **Autorun** from the **View - Events** menu, and then save the spreadsheet (select **Save** from the **File** menu).

The next time you open the **Exp.sta** datafile, the new event handler will automatically be executed, and you will not be able to double-click on a cell in the spreadsheet. To "disconnect" the event handler, select **Autorun** again from the **View - Events** menu to clear the command.

Example: Responding to Application-Level Events

Application-level events can be used to customize behavior of all *STATISTICA* documents. To create a similar effect as that described in the previous example (for the spreadsheet), follow these steps:

Select **View Code** from the **Tools - Macro - Application Events** menu. In the **Document Events** window, select **Document** in the **Object** box and select **SpreadsheetBeforeDoubleClick** in the **Proc** box. Then change the event procedure like this:

```
Private Sub Document_SpreadsheetBeforeDoubleClick( _
       ByVal Spreadsheet As Object, _
       ByVal Flags As Long, _
       ByVal VarNo As Long, _
       ByVal CaseNo As Long, _
       Cancel As Boolean)
            MsgBox "Double-click not allowed"
            Cancel = True
End Sub
```

Next, run this event handler, and select **Autorun** from the **Tools - Macro - Application Events** menu. Then, close *STATISTICA* and restart it. Any time you double-click on a cell in any spreadsheet, the *"Double-click not allowed"* message will be displayed. To "disconnect" this event handler, clear the **Autorun** command on the **Tools - Macro - Application Events** menu.

Supported Events

STATISTICA supports event handlers for the application (as demonstrated in *Example: Responding to Application-Level Events*, see above) and all objects, i.e., spreadsheets, graphs, workbooks, and reports. You can review the list of available events that can be customized by reviewing the contents of the **Proc** box in the **Document Events** dialog.

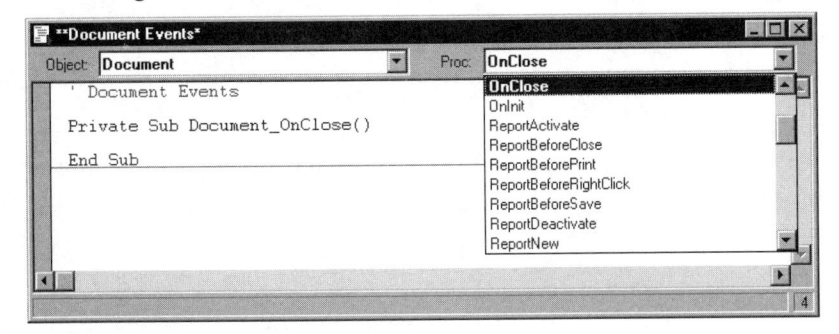

```
**Document Events**
Object: Document          Proc: OnClose
    ' Document Events            OnClose
                                 OnInit
    Private Sub Document_OnClose()
                                 ReportActivate
    End Sub                      ReportBeforeClose
                                 ReportBeforePrint
                                 ReportBeforeRightClick
                                 ReportBeforeSave
                                 ReportDeactivate
                                 ReportNew
```

Practically all events related to the opening, printing, saving, closing, or editing of documents or applications can be "intercepted" by custom event handlers. Since the entire functionality of the *STATISTICA* Visual Basic programming environment can be used to customize these event handlers, it is easy to see how completely customized versions of the *STATISTICA* application can be produced that will alter the "basic behavior" of various options and user actions and, thus, optimize the program for particular routine tasks or particular operators with limited access permissions, etc.

Customizing Toolbars and Menus via *STATISTICA* Visual Basic

STATISTICA provides full access to all of its customization options via *STATISTICA* Visual Basic functions. A detailed discussion of the techniques available for customizing toolbars, menus, etc., and how to use these methods to fully customize or expand *STATISTICA* is beyond the scope of this introduction. There are many excellent books available that discuss in great detail the *CommandBars* object and how to use it in (Microsoft) Windows applications (e.g., to customize Microsoft Office applications). In general, customizing toolbars, menus, pop-up menus, etc. is useful in order to:

- Provide quick access to frequently used options or methods;
- Program highly customized events; e.g., to customize the shortcut menu when you right-click on certain objects by adding macro programs or links to other applications to that menu;

CHAPTER 7: CUSTOMIZING *STATISTICA* WITH SVB

- Program fully customized versions of *STATISTICA* dedicated to perform only a few predefined tasks (e.g., a version for data entry only);

- Add new options to *STATISTICA* developed by other vendors; those options could be invoked from a *STATISTICA* Visual Basic macro program if they are accessible as reference libraries.

For details on how to customize toolbars and menus via *STATISTICA* Visual Basic, refer to the *Electronic Manual* or printed documentation.

CHAPTER 8

MATRIX AND STATISTICAL FUNCTION LIBRARIES

Include File: STB.svx 83

A Simple Example: Inverting a Matrix 84

CHAPTER 8

MATRIX AND STATISTICAL FUNCTION LIBRARIES

STATISTICA Visual Basic contains a large number of designated matrix and statistical functions that make the SVB environment ideal for prototyping algorithms or for developing custom statistical procedures. The matrix and statistical functions are documented in detail in *STATISTICA Matrix Function Library* in the *Electronic Manual*.

One major advantage of using the *STATISTICA* library of matrix functions instead of writing these functions "by hand" in Visual Basic is that the former will evaluate much faster. For example, when you want to invert large matrices, the `MatrixInverse` function will perform the actual matrix inversion using the highly optimized (compiled) algorithms of *STATISTICA*.

Accessing data in spreadsheets. In many cases, you may want to access the data in a spreadsheet via the matrix functions. You can of course use the spreadsheet `.Value(i,j)` property to retrieve and set individual values in a data spreadsheet. In addition, three specialized spreadsheet properties provide more efficient methods for transferring data to and from matrices suitable for use with *STATISTICA* matrix functions: `.Data, VData(i), and CData(i)`. The `.Data` property will transfer all data in a spreadsheet to a two-dimensional array (matrix), and back; the `.VData(i)` property will transfer only specific variables or columns (the *i*'th column); the `CData(i)` property will transfer only specific cases or rows (the *i*'th row). These properties are illustrated in examples below.

Include File: STB.svx

To provide convenient access to the matrix functions without requiring you to pass explicitly the dimensions of the arrays that are being passed, *STATISTICA* includes a file with function interfaces: **STB.svx**. You may want to open this file to review the functions and how they provide simplified access to the actual matrix library. It is recommended that you routinely

CHAPTER 8: MATRIX AND STATISTICAL FUNCTIONS

include that file at the beginning of your program when you intend to use functions from the *STATISTICA* libraries of matrix and statistical functions.

A Simple Example: Inverting a Matrix

The following example illustrates how to use the functions in the *STATISTICA* library of matrix functions to perform some basic matrix operations, namely, matrix inversion and matrix multiplication:

```
' The following line is very important!
' It contains the "interfaces" to the STATISTICA
' Matrix and Statistical Function libraries. Note
' that we are using the "star" (*) convention to
' direct the program to "look for" file STB.svx
' in the current installation of STATISTICA; this
' ensures portability of code from one machine
' (installation of STATISTICA) to another.
'$include: "*STB.svx"
' The next statement will cause all arrays to be
' declared by default as 1-referenced, i.e., the
' first element in an array will be element 1, not 0.
Option Base 1
Sub Main
  Dim a(4,4) As Double, ainv(4,4) As Double
  Dim i As Long, j As Long
  For i=1 To 4
   For j=1 To 4
    a(i,j)=Rnd(10)
   Next j
  Next i
  MatrixDisplay (a, "Original Matrix A")
  MatrixInverse(a,ainv)
  MatrixDisplay (ainv, "A, Inverse")
End Sub
```

A few things should be noted. First, be sure to include file **STB.svx** before making the calls to the matrix library, as shown above; otherwise, the program will not run. When an SVB program contains a reference to an external class or object module, it may use the asterisk convention in which case during execution the program will "look in" the global macro directory (and other standard *STATISTICA* directories) for those files. Second, the statement `Option Base 1` causes all arrays to be created, by default, as 1- referenced arrays. In other words, the first element in arrays can be referenced as element number 1, the second as number 2, etc. If the `Option Base 1` statement is omitted, then by default, arrays are declared as 0-

referenced, i.e., the first element in an array is element 0 (zero), the second element is element 1, etc. While it is not necessary that the arrays used in the calls to matrix functions (via the interfaces in **STB.svx**) are 1-referenced, it can make writing programs using these functions a little less confusing.

CHAPTER 9

INTRODUCTORY EXAMPLES

Displaying a Simple Message Box 89
Making a Spreadsheet and Filling It with
    Random Numbers 90
Displaying a Progress Bar 90
Making a Histogram with Normal Fit 91
Placing Results in Workbooks, Reports, Etc., via
    the RouteOutput Method 92
Sending Results to a Report Window 93

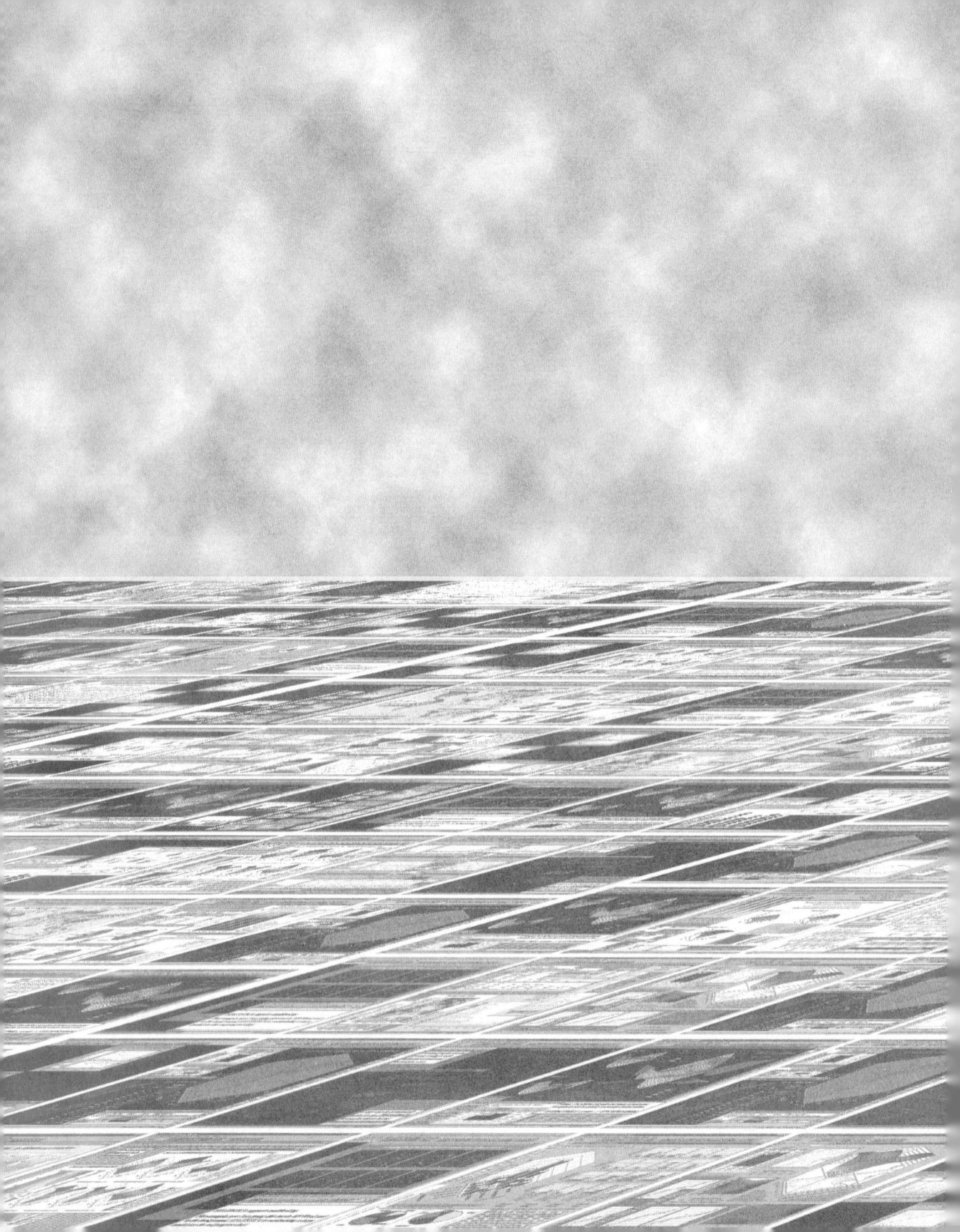

CHAPTER 9

INTRODUCTORY EXAMPLES

This section includes a number of very short and simple SVB programs to illustrate specific "tasks" and how they can be accomplished.

Displaying a Simple Message Box

This program will cause a simple message box to be displayed.

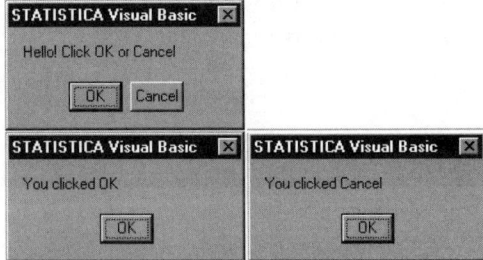

```
Sub Main
' NOTE: The next statement brings up a message box.
    If MsgBox ("Hello! Click OK or Cancel", _
        vbOkCancel)=vbOK Then
      MsgBox "You clicked OK"
    Else
      MsgBox "You clicked Cancel"
    End If
End Sub
```

When you run the program you will see an initial message box and, depending on whether you click **OK** or **Cancel**, a second message box indicated by your choice (see above).

Making a Spreadsheet and Filling It with Random Numbers

This example creates a new spreadsheet and fills it with random numbers. The first column will be filled with uniform random numbers; the second column will be filled with normal random numbers.

```
Option Base 1
Sub Main
 Dim n As Long,i As Long
 n=1000
 ' Create and dimension the spreadsheet object
 Dim s As New Spreadsheet
 ' Set the size of the object
 s.SetSize(n,2)
 ' Assign the random values
 For i=1 To n
  s.Value(i,1)=Rnd(1)
  s.Value(i,2)=RndNormal(1)
 Next i
 ' Set the variable names
 s.VariableName(1)="Uniform"
 s.VariableName(2)="Normal"
 s.Visible=True
End Sub
```

Displaying a Progress Bar

It is sometimes desirable to indicate the progress of lengthy sequential computations by displaying a progress bar, the same kind of progress bar that is used throughout *STATISTICA* when large datafiles are analyzed. Here is a program that implements the progress bar in the random number generator program:

```
Option Base 1
Sub Main
 Dim n As Long,i As Long
 n=1000
 ' Create and dimension the spreadsheet object
 Dim s As New Spreadsheet
 ' Set the size of the object
 s.SetSize(n,2)
 ' Set up a progress bar
 Dim pb As ProgressBar
```

```
  Set pb = AddProgressBar("Generating random numbers", 1, n)
' Assign the random values
 For i=1 To n
' Update the progress bar
  pb.CurrentCounter = i
  s.Value(i,1)=Rnd(1)
  s.Value(i,2)=RndNormal(1)
 Next i
' Close the progress bar
 Set pb = Nothing
' Set the variable names
 s.VariableName(1)="Uniform"
 s.VariableName(2)="Normal"
 s.Visible=True
End Sub
```

Making a Histogram with Normal Fit

The next example illustrates how to make a simple histogram from some data generated in the program; specifically two histograms will be created for the data generated in the previous example.

```
Option Base 1
Sub Main
Dim n As Long
 n=1000
 Dim s As New Spreadsheet
' Create the random numbers in spreadsheet s
 ComputeRandomNumbers s, n
' Create the histogram from the numbers in s
 CreateHistograms s
End Sub

Sub ComputeRandomNumbers (s As Spreadsheet, n As Long)
 Dim i As Long
 ReDim x(n,2) As Double
  s.SetSize(n,2)
  For i=1 To n
   x(i,1)=Rnd(1)
   x(i,2)=RndNormal(1)
  Next i
  s.Data=x
  s.VariableName(1)="Uniform"
  s.VariableName(2)="Normal"
End Sub
```

CHAPTER 9: INTRODUCTORY EXAMPLES

```
Sub CreateHistograms (s As Spreadsheet)
' NOTE: The following code was created by modifying (simplifying)
' the code recorded in a simple analysis macro.
  Dim newanalysis As Analysis
  Set newanalysis = Analysis (sc2dHistograms, s)
  With newanalysis.Dialog
     .Variables = "1 2 | "
     .GraphType = scHistgoramRegularPlot
  End With
  newanalysis.Dialog.Graphs.Visible = True
End Sub
```

Note that this example program calls a subroutine to compute the data for the input spreadsheet and another to create the histograms.

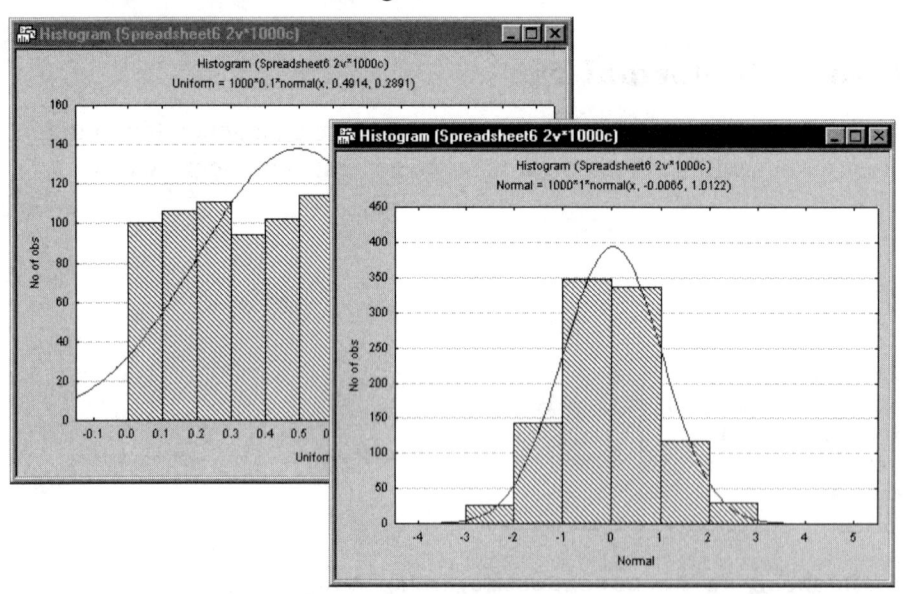

The histogram is created via a *STATISTICA* analysis, and the code as shown above was created by recording a simple analysis in an analysis macro, and then simplifying the recorded program (i.e., mostly deleting the explicit recording of various default options).

Placing Results in Workbooks, Reports, Etc., via the RouteOutput Method

You can send results graphs or spreadsheets from your SVB programs to the same "locations" (e.g., workbooks, reports) as all other results from your analyses. The `RouteOutput` method will create the respective results graph or spreadsheet in a workbook or report if your *STATISTICA* is currently configured accordingly. See also the description of the `RouteOutput` method page 41) for details. For example, in the previous example, in `Sub CreateHistograms` replace the line:

```
newanalysis.Dialog.Graphs.Visible = True
```

with:

```
newanalysis.RouteOutput(newanalysis.Dialog.Graphs).Visible = True
```

Now the histograms will be created in the default location(s) as specified in the **Output Manager** dialog. You can also refer to the *Advanced Examples* (page 97) to see how to create workbooks from within an SVB program, and how to send explicitly results spreadsheets or graphs to workbooks.

Sending Results to a Report Window

The `RouteOutput` method (see also page 41) will create results in the same channel as all other output, depending on your current configuration of *STATISTICA* (in the **Output Manager** dialog). You can also send text, graphs, etc. directly to a report window. Here is an example program that shows how to do that. Note that routine `ComputeRandomNumbers` that was used in the previous examples is not repeated below.

```
Option Base 1
Sub Main
Dim g(2) As Graph
Dim n As Long
Dim s As New Spreadsheet
' Create a new report window.
Dim r As New Report
 n=1000
' Create the random numbers in spreadsheet s
 ComputeRandomNumbers s, n
' Create the histogram from the numbers in s
 CreateHistograms s, g
' Add the following text
 r.SelectionText="This is the graph for the Uniform variate:" _
   +vbCrLf+vbCrLf
' Add the first graph to the report
 r.SelectionObject=g(1)
' Add two <cr>-<lf> (new lines)
 r.SelectionText=vbCrLf+vbCrLf
 r.SelectionText="This is the graph for the Normal variate:" _
   +vbCrLf+vbCrLf
' Add the second graph to the report
```

CHAPTER 9: INTRODUCTORY EXAMPLES

```
      r.SelectionObject=g(2)
      r.SelectionText=vbCrLf+vbCrLf
    ' Make the report visible
      r.Visible=True
    End Sub
    Sub CreateHistograms (s As Spreadsheet, g() As Graph)
    ' NOTE: The following code was created by modifying (simplifying)
    ' the code recorded in a simple analysis macro.
      Dim newanalysis As Analysis
      Set newanalysis = Analysis (sc2dHistograms, s)
      newanalysis.Dialog.Variables = "1 2 | "
      Set g(1)=newanalysis.Dialog.Graphs(1)
      Set g(2)=newanalysis.Dialog.Graphs(2)
    End Sub
```

When you run this program, the following results report window will be created; note the two text titles that are created by the program and placed into the same report with the graph objects.

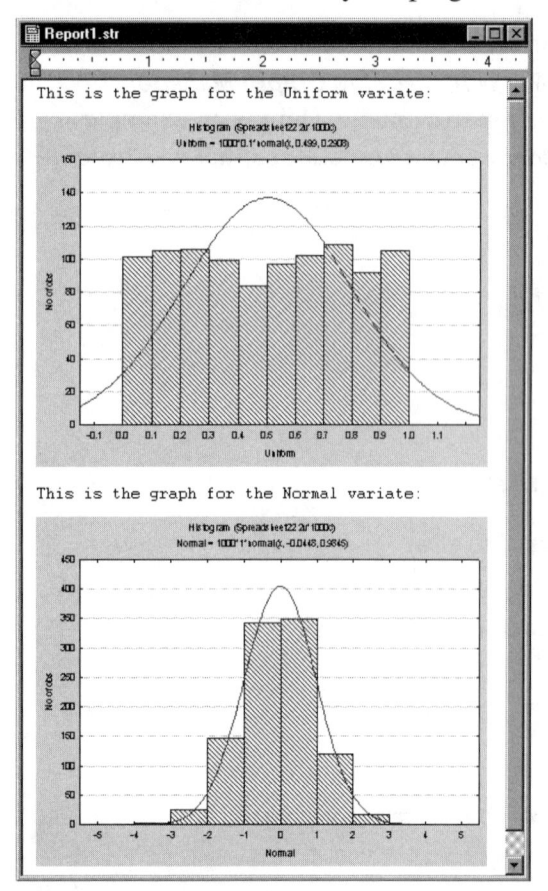

94 – *STATISTICA* Visual Basic Primer

CHAPTER 10

ADVANCED EXAMPLES

Creating and Customizing Graph Objects 98
Creating a Cell-Function Spreadsheet 104
SVB Program for a By-Group Analysis 107

CHAPTER 10

ADVANCED EXAMPLES

The examples in this section show how you can use SVB programs to add new functionality to *STATISTICA*. Specifically, the first set of example programs illustrates how to navigate the object model for *STATISTICA* graphs, i.e., how to create a blank graph, set data into that graph, and apply various modifications such as custom text, arrows, etc. The general principles illustrated in these examples will apply to other *STATISTICA* objects as well, and after reviewing these examples, you will find it much easier to navigate the object models for spreadsheets, reports, workbooks, etc.

The remaining two sections (programs) illustrate how to write useful applications that customize events and how to build a complete custom program using the SVB tools. Specifically, the first program (page 104) describes how to customize document-level events to program an "Excel-like" spreadsheet that will automatically update whenever one of the input values (into a budget) are changed. The second example (page 107) demonstrates how to enhance a recorded macro to perform an advanced analysis-by-group (multiple regression analyses broken down by one or more categorical variables) by adding a dialog and some custom handling of output.

To summarize, these examples programs demonstrate:

- How to create customized graphs, and manage graphics fonts, extra objects, etc;
- How to design and service dialogs;
- How to call functions to accept variable lists;
- How to perform error trapping;
- How to manipulate strings;
- How to create strings to be used for case selection conditions;
- How to pass variable selections from custom dialogs to analyses (**Multiple Regression**);

CHAPTER 10: ADVANCED EXAMPLES

- How to move spreadsheets and graphs into workbooks;
- How to change titles of workbook items;
- How to control document-level events;
- How to write protect cells in a data spreadsheet.

What is not shown in these examples. *STATISTICA* Visual Basic is a very powerful general programming language that will allow you to customize and control virtually all functionality of *STATISTICA*. Hence, it is impossible to illustrate all useful features in just a few examples. Some of the important (often used) features of SVB that are not illustrated here:

- How to extensively customize spreadsheets, printed reports, etc. (fonts, colors, patterns, etc.);
- How to use third-party functions (type libraries);
- How to build *STATISTICA* into other applications (e.g., Microsoft Excel);
- How to access and query external databases.

Note that example SVB programs that illustrate some of these functions are included in the library of SVB programs distributed with *STATISTICA*.

Creating and Customizing Graph Objects

The following examples demonstrate how to create an empty graph, set data into the graph, and then add various customizations.

Creating a blank graph. Let us begin by creating a simple graph object. Create a new macro program and enter the program instructions shown below; after you run the program a blank graph will be displayed.

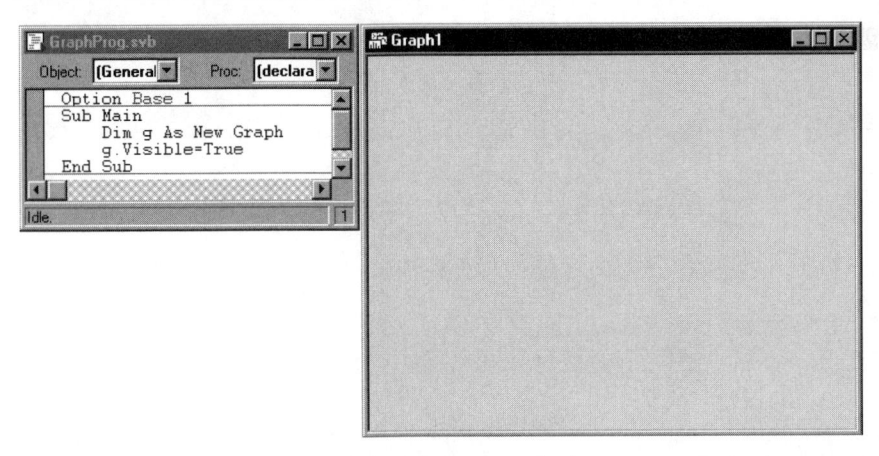

```
Option Base 1
Sub Main
    Dim g As New Graph
    g.Visible=True
End Sub
```

At this point variable **g** will hold the graph; however, there is no content (and content area), so the next thing to do is to create a graph object to hold any data, customizations, etc. To do so, type in and run the following program:

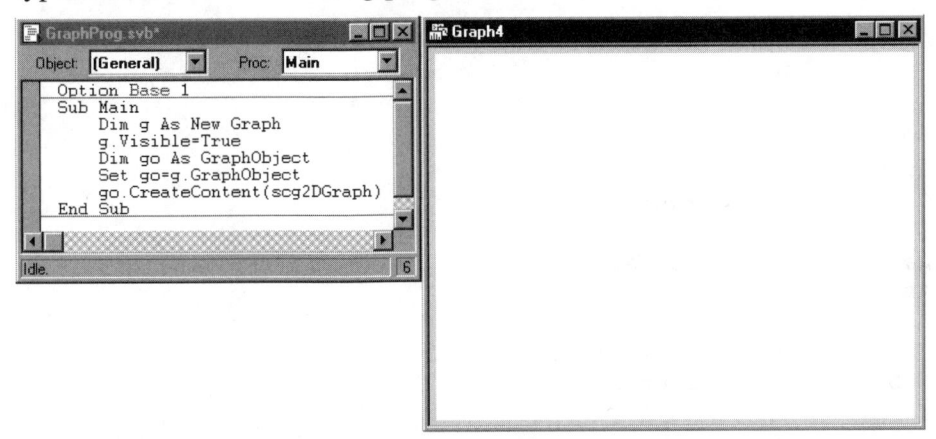

```
Option Base 1
Sub Main
    Dim g As New Graph
    g.Visible=True
    Dim go As GraphObject
    Set go=g.GraphObject
    go.CreateContent(scg2DGraph)
End Sub
```

We have now created a blank 2D graph and have the object with all its properties and methods in variable **go**. As mentioned in various places throughout this documentation, to quickly review the object model for graphs, you can display the **Object Browser** and select the *STATISTICAGraphics* library.

CHAPTER 10: ADVANCED EXAMPLES

Setting data into the graph. Next, let us set data into the blank graph. To do so, type in the following program:

```
Option Base 1
Sub Main
    Dim g As New Graph
    Dim go As GraphObject
    Set go=g.GraphObject
    go.CreateContent(scg2DGraph)
' Create data:
' Henon Strange Attractor; see Welstead, 1994, Neural Networks.
    Dim Size As Long
    Size=10000
    ReDim h1(Size) As Double, h2(Size) As Double
    Dim hh1 As Double, hh2 As Double
    hh1=0
    hh2=0
    For i=1 To Size
      h1(i)=1 + hh2 - 0.1*hh1*hh1
      h2(i)=.9998*hh1
      hh1=h1(i)
      hh2=h2(i)
    Next i
' Now set the data in the 2D Graph, as a scatterplot:
' First add a plot:
    Dim l2 As Layout2D
    Set l2=go.Content
' Variable l2 contains the 2D Layout object; next
' retrieve the plots in the current graph.
    Dim ps As Plots2D
```

```
        Set ps=12.Plots
' Create a new 2D simple plot via the Add method.
        Dim p2d As Plot2D
        Set p2d=ps.Add(scgSimplePlot,Size)
' Now retrieve the variables in the first plot, i.e. the one
' we just created.
        Dim Vx As Variable, Vy As Variable
        Set Vx=p2d.Variable(1)
        Set Vy=p2d.Variable(2)
' Finally, set the new values.
        For i=1 To Size
          Vx.Value(i)=h1(i)
          Vy.Value(i)=h2(i)
        Next i
' Make the graph visible
        g.Visible=True
End Sub
```

Here is the graph that will be created:

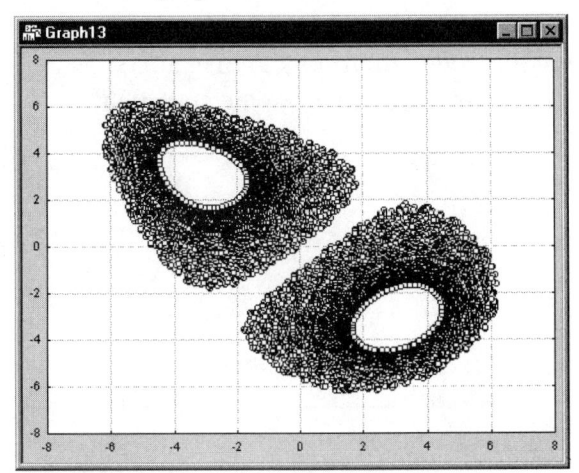

Adding a title to the graph. Next, let us write a subroutine to add a title to the graph. To do so, right before the statement `g.Visible=True` add a call to a new subroutine, and define the subroutine as shown below:

```
        ...
        ...
        SetTitleOfMyGraph g,"Henon Strange Attractor", "Arial Black", 15
        g.Visible=True
End Sub

Sub SetTitleOfMyGraph(g As Graph,MyTitle As String, _
```

CHAPTER 10: ADVANCED EXAMPLES

```
   FontName As String, FontSize As Long)
    Dim t As Titles
    Set t=g.Titles
    Dim t1 As Title
' Create a new main title.
    Set t1=t.Add(scgMainTitle,MyTitle)
    t1.Font.Size=FontSize
    t1.Font.FaceName=FontName
End Sub
```

Note how we not only have created a title, but also modified the default font and font size for the title. Also, for illustration purposes, this code could be written much more "densely" by not explicitly defining the respective objects; so an alternate way to write this routine would be:

```
Sub SetTitleOfMyGraph(g As Graph,MyTitle As String, FontName As
String, FontSize As Long)
    g.Titles.Add(scgMainTitle,MyTitle)
    g.Titles.Item(1).Font.Size=FontSize
    g.Titles.Item(1).Font.FaceName=FontName
End Sub
```

Changing the scales. You can now add the following code (subroutines) to change the scaling for the axes and the font that is used for the axis labels. Again, before the statement `g.Visible=True` add the calls to the functions as shown below:

```
    ...
    ...
        ChangeAxisScaling 12.Axes.XAxis, -8.0, 1.0, 8.0, _
            "Arial Black", 12
        ChangeAxisScaling 12.Axes.YAxis, -8.0, 1.0, 8.0, _
            "Arial Black", 12
        g.Visible=True
End Sub

Sub ChangeAxisScaling (x As Axis2D, _
     xmin As Double, xstep As Double, xmax As Double, _
     FontName As String, FontSize As Long)
  x.RangeMode=scgManualRange
  x.SetManualRange(xmin,xmax)
  x.StepMode=scgManualStep
  x.StepSize=xstep
  x.Font.Size=FontSize
  x.Font.Face.FaceName=FontName
End Sub
```

Extra objects: Adding custom text. Next, let us write a subroutine to add a brief description of what is being plotted as a custom text. Again, we will make a subroutine, this time to place some custom text in a particular location in the graph.

```
...
...
  AddCustomTextToMyGraph g, _
     "Compute Recursively:" + vbCrLf + _
     "   x(i)=1+y(i-1)-.1*x(i-1)^2" + vbCrLf + _
     "   y(i)=.9998*x(i-1) ", _
     7, 7, "Arial Black", 12
   g.Visible=True
End Sub

Sub AddCustomTextToMyGraph ( g As Graph, _
   t As String, x As Double, y As Double, _
   FontName As String, FontSize As Long)
   Dim tob As TextObject
   Set tob=g.ExtraObjects.AddDynamicText(t,x,y)
   tob.Parameters.AnchorPosition=2
   tob.Text.Font.FaceName=FontName
   tob.Text.Font.Size=FontSize
End Sub
```

Depending on your current system defaults, the final graph will look like this:

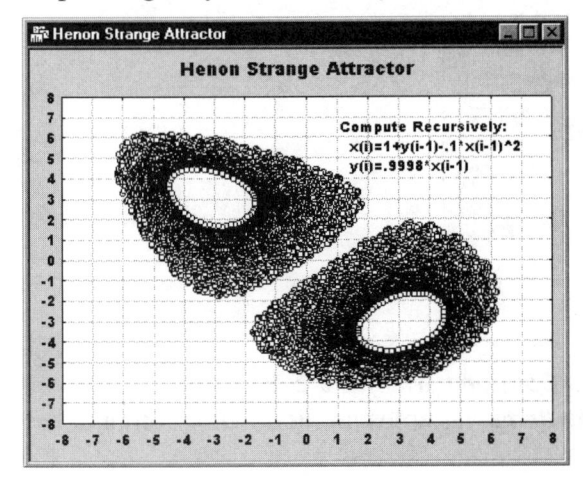

Note that the `Graph.ExtraObjects` object contains methods to add not only text but also (poly-) lines, arrows, shapes, and other graphs, i.e., to embed those custom (extra-) objects into your graph. Thus, the *STATISTICA* graphics library provides a comprehensive tool to build highly customized graphical displays.

CHAPTER 10: ADVANCED EXAMPLES

Creating a Cell-Function Spreadsheet

One of the most basic functions of the designated spreadsheet software (such as Microsoft Excel) is to automatically recompute cells in the datafile when any of the input data (cells) are changed. For example, you can set up a complex budget for a project so that when you change the values for particular budget items, the entire budget will be recomputed based on the newly supplied values.

The same functionality can be programmed into *STATISTICA* Spreadsheets by attaching an SVB macro to certain spreadsheet events as demonstrated in this example. Note that practically all spreadsheet (and other) events can be customized, thus providing the tools to build very sophisticated and highly customized automated data operations right "into" the spreadsheet.

How is the macro created? First, create a datafile and set up the necessary cells. Datafile *CellFunctionDemo.sta* (available in the Examples\Macros directory that can be found in the directory in which you installed *STATISTICA*) contains prices for various items on a holiday shopping list.

Christmas Shopping List	This example demonstrates using variable functions and the spreadsheet-level event "DataChanged()" to update both a variable and a single cell. Note that the "Final Item Cost" (defined with a spreadsheet formula) and the "Total Cost of All Items" (defined by a STATISTICA Visual Basic macro) columns are protected using the custom event handling (and cannot be changed); both columns will recalculate automatically as you change the "Item Cost" and/or the "Coupon" values. To review or modify the code of the program attached to this spreadsheet use the menu View / Events / View Code.			
	Item Cost	Coupon	Final Item Cost	Total Costs of All Items
Board Games	$75.00	$10.00	$65.00	
Movies	$250.00	$5.00	$245.00	
DVD Player	$200.00	$10.00	$190.00	
Computer Software	$140.00	$8.00	$132.00	
Video Games	$100.00	$10.00	$90.00	
Clothes	$80.00	No Coupon	$80.00	
Gift Certificates	$70.00	No Coupon	$70.00	
Books	$90.00	$5.00	$85.00	$957.00

Note that in the spreadsheet shown, the *Item Cost* and *Coupon* variables are data input variables, and the *Final Item Cost* and *Total Cost of All Items* variables are derived or computed variables.

Entering the computations for derived cells (programming the DataChanged event). After entering the basic information, select *View Code* from the *View - Events* menu. This displays the SVB program editor for document-level events (i.e., events that apply to the newly created spreadsheet document). Select *Document* in the *Object*

104 – *STATISTICA* Visual Basic Primer

box of the SVB editor (**Document Events** window); select the **DataChanged** event in the **Proc** box. Shown below is the appearance of the SVB editor after the `DataChanged` subroutine has been created in the editor.

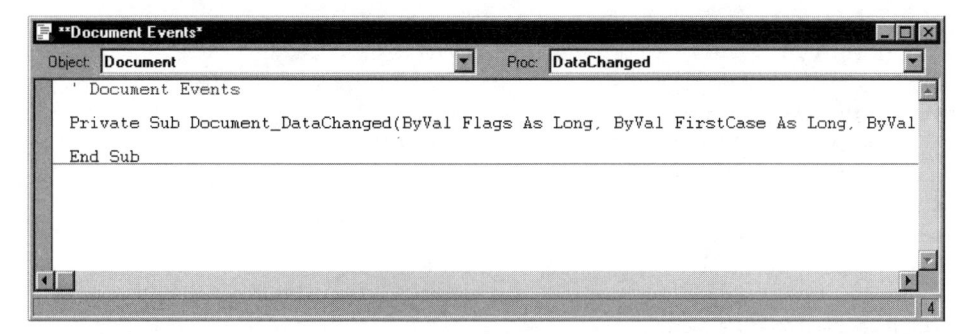

```
**Document Events**
Object: Document                              Proc: DataChanged
' Document Events
Private Sub Document_DataChanged(ByVal Flags As Long, ByVal FirstCase As Long, ByVal
End Sub
```

Now type the following program into the SVB editor (you can, of course, omit the comments).

```
Private Sub Document_DataChanged(ByVal Flags As Long, _
   ByVal FirstCase As Long, ByVal FirstVar As Long, _
   ByVal LastCase As Long, ByVal LastVar  As Long, _
   ByVal bLast As Boolean)
' Only process the data if there was a change in
' the data area of the spreadsheet.
   If (Flags And scNotifyDCData) Then
   If FirstVar = 3 Or FirstVar = 4 Or _
    LastVar = 4 Or LastVar  = 4 Then
    MsgBox "These are derived fields And they cannot be edited."
   End If

   Const V1 As Long = 1
   Const V2 As Long = 2
   Const VDest As Long = 3
   Const VResult As Long = 4
   Dim s As Spreadsheet

   Set s = ActiveSpreadsheet
' We need to recalculate first to update the final cost of each item;
' we only need to do so for the range of cases that have changed.
   Dim j As Long
   For j = FirstCase To LastCase
    Dim x1 As Double
    Dim x2 As Double
' If the source data is missing data, then substitute 0.
    If (s.MissingData(j, V1)) Then
     x1 = 0
    Else
     x1 = s.Value(j,V1)
```

```
            End If
' If the source data is missing data, then substitute 0.
            If (s.MissingData(j,V2)) Then
              x2 = 0
            Else
              x2 = s.Value(j,V2)
            End If
' Calculate new destination variable
             s.Value(j, VDest) = x1 - x2
          Next j
         Dim i As Long
         Dim TotalVal As Double
' Iterate through each cell in variable 3 and add it to
' TotalVal.
         For i = 1 To s.NumberOfCases
             TotalVal = TotalVal + s.Value(i, VDest)
         Next i
' Update the cell's value to reflect total cost of all items.
         s.Value(s.NumberOfCases, VResult) = TotalVal
       End If
End Sub
```

This macro defines the computations for the cells in the spreadsheet that will be performed every time the data in the input variables are changed.

Write protecting the derived cells. Also, we want to make sure that certain cells are protected, i.e., users should not be able to type values into the cells that are derived (by computation from other cells). Some of that protection is already implemented in the macro shown above, which checks whether or not the user attempted to type a value into a derived cell. For this example, let us also "catch" the *BeforeDoubleClick* event for the cells in the third and fourth variables of our example spreadsheet. Select the **BeforeDoubleClick** event in the **Proc** field of the SVB editor, and then enter the code as shown below:

```
    Private Sub Document_BeforeDoubleClick(ByVal Flags As Long, ByVal
    CaseNo As Long, ByVal VarNo As Long, Cancel As Boolean)
        If VarNo = 3 Or VarNo = 4 Then
    MsgBox   "These are derived fields and they cannot be edited."
           Cancel = True
        End If
    End Sub
```

Saving the spreadsheet and AutoRun. Finally, before saving the macro and the datafile, click on the data spreadsheet once more, and select **Autorun** from the **View – Events** menu.

This will cause the new macro to run automatically every time you open the data spreadsheet. Next, save the spreadsheet and run the macro. You are now ready to compute your holiday shopping budget using your customized spreadsheet. If you try to "cheat" by double-clicking on one of the computed fields to enter a (lower) total value, a message will be displayed.

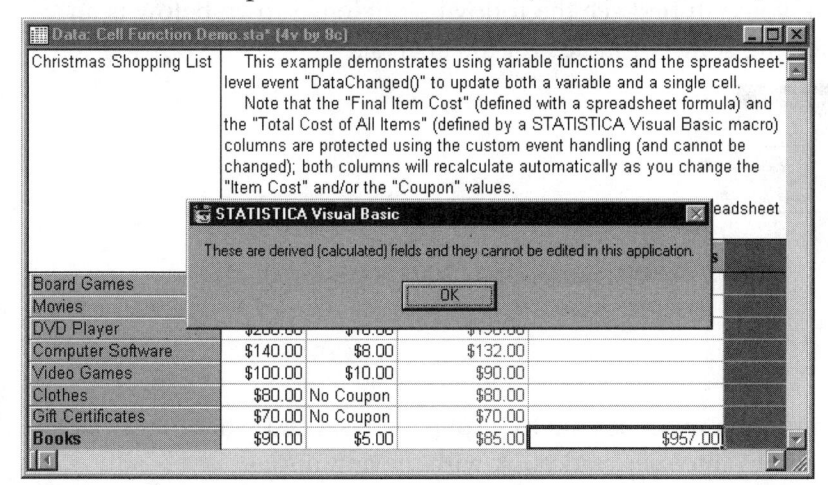

This simple example illustrates how you could build very sophisticated "cost models" that can also include dialogs, automatic analyses with the *STATISTICA* statistical or graphics functions, or any of the more than 10,000 automation functions available in the *STATISTICA* system, thus vastly expanding the functionality of ordinary spreadsheets.

SVB Program for a By-Group Analysis

The following more complex program illustrates how to perform a by-group analysis. In this specific instance, the program will display a dialog (created in SVB) to accept the predictor and dependent variables for a multiple regression analysis and selections of desired output from that analysis. The user can also specify one or more categorical (grouping) variables, and the program will then perform the requested multiple regression analyses, broken down by the unique combinations of values in the categorical variables. This is accomplished by creating case selection conditions (merging with any existing case selection conditions) that select for each (within-group) analysis only the respective cases that belong to the respective combination of categorical values in the categorical (grouping) variables.

CHAPTER 10: ADVANCED EXAMPLES

First, open the datafile that is to be used for the analysis; for example, you might use the example datafile *Exp.Sta,* which contains several continuous and categorical variables (e.g., a categorical variable *Gender*). Then create a new SVB program macro and enter the program shown below; this program is also included as an example with your *STATISTICA* program (*Multiple Regression By-Group.svb*), so you may just want to open that file and try running the macro.

When you run this macro program, you will first see the following dialog. Shown below is an illustration of that dialog after selecting some variables for the analysis.

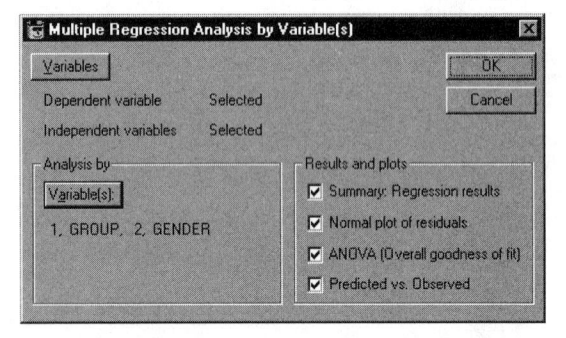

After clicking the *OK* button, the program produces the by-group analysis and places the selected results spreadsheets and graphs into one workbook with the appropriate labels.

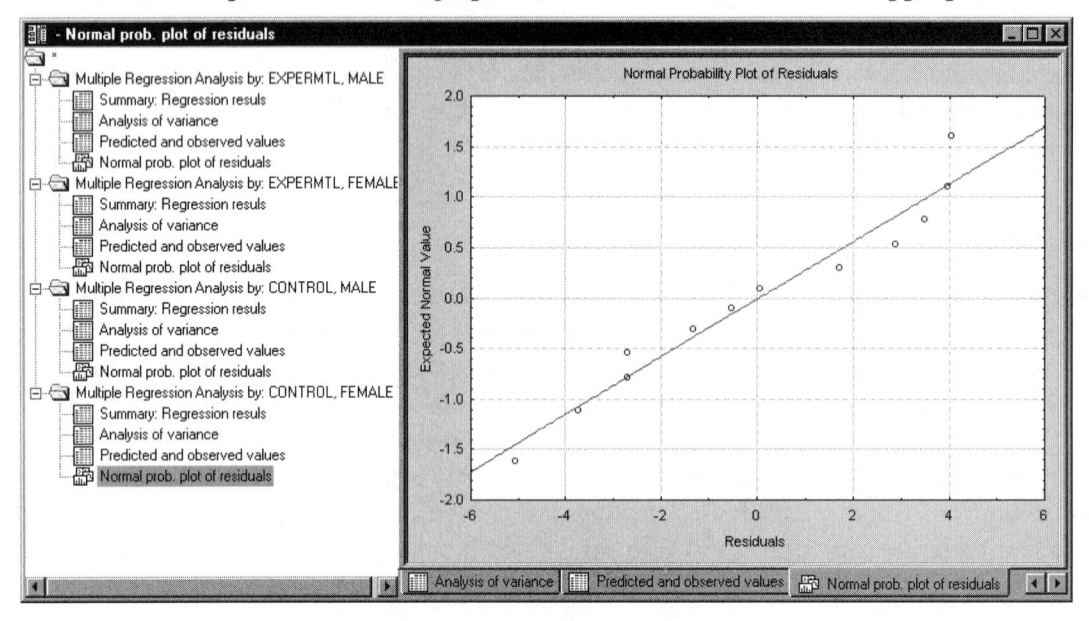

Here is the annotated listing of the program:

108 – *STATISTICA* Visual Basic Primer

```vb
'This program will perform a multiple regression analysis by groups. The
'user will be prompted to specify the variables for the multiple regression,
'and one or more categorical variables for by-group analysis. The program
'will then perform the requested regression analysis for each combination of
'the values of the categorical (grouping) variables (e.g., Male-
'Experimental, Male-Control, Female-Experimental...). The by-group analysis
'is performed by assembling case selection conditions that will select for
'each analysis only those cases that belong to the respective combination
'of values in the selected categorical variables. If the input datafile
'has any active pre-set case selection conditions "connected" already with
'the analysis, then these will be merged with the ones assembled in this
'program.

        Option Base 1
        Dim Inlist1 As Long
        Dim Inlist2 As Long
        Dim Inlist3 As Long
        Dim Inlist4 As Long
        Dim Inlist5 As Long
        Public Varlist1() As Integer
        Public Varlist2() As Integer
        Public Varlist3() As Integer
        Public Varlist4() As Integer
        Dim Varlist5() As Long
        Dim List1Title As String
        Dim List2Title As String
        Dim List3Title As String
        Dim List4Title As String
        Dim VarCodeNumber() As Long
        Dim Nvars As Long
        Dim Ncases As Long
        Dim NumberAnalysisVarLists As String
        Dim NumberCriterionVars As Long
        Dim AdditionalCaseSelectCond() As String
        Public ResultsWorkbook As Workbook
        Dim FolderCaseName() As String
        Public Folder As WorkbookItem
        Public CurrentDataSet As Spreadsheet
        Dim ResultOption1 As String
        Dim ResultOption2 As String
        Dim ResultOption3 As String
        Dim ResultOption4 As String

        Sub Main
        'The comments in the following code use the term (i) "case-selection-
        'variable(s)" for the variable(s) by which the analysis has to be
```

```
'performed, i.e., the variables whose categories will be used as
'criterions for selecting the cases which will be included in the
'computations, (ii) "analysis-variables" for the variables which will
'actually participate in the computations and display of the results.
Dim FolderTitle As String
'USER INPUT SECTION BEGINS
'User specifies the number of variables (exclusive of the case-
'selection variable(s)).
NumberAnalysisVarLists = "2"
'User specifies the titles of the list(s) of analysis-variable(s).
List1Title = "Dependent variables"
List2Title = "Independent variables"
'User specifies the folder title.
FolderTitle = "Multiple Regression Analysis"
'USER INPUT SECTION ENDS.
Dim ResSpreadsheet As Spreadsheet
Dim Resgraph As Graph
Dim MaxNoCriterionVars As Long
Dim MaxNoCategories As Long
Dim CriterionVariable As String
Dim NewSelectionCond As String
Dim NumberOfCategCombns As Integer
Dim i As Integer
Dim Ret As Integer
Dim ifault As Integer
'Set the source spreadsheet for the input data from which the number
'of cases and the number of variables will be determined and the
'current analysis performed.
Set CurrentDataSet = ActiveDataSet
'Determine the number of cases and the number of variables in the
'current data set.
Nvars = CurrentDataSet.NumberOfVariables
Ncases = CurrentDataSet.NumberOfCases
'Set the maximum number of case-selection variables, and the maximum
'number of categories for each variable, that are permissible in the
'current analysis.
MaxNoCriterionVars = 50
MaxNoCategories = 50
'Dynamically, dimension the analysis-variables and the case-selection
'variables lists.
ReDim Varlist1(1 To Nvars)
ReDim Varlist2(1 To Nvars)
ReDim Varlist3(1 To Nvars)
ReDim Varlist4(1 To Nvars)
ReDim Varlist5(1 To Nvars)
'Initialize the four lists for analysis-variables as well as for the
```

```
'case-selection variables list.
Varlist1(1) =0
Inlist1=0
Varlist2(1) =0
Inlist2 =0
Varlist3(1)=0
Inlist3 =0
Varlist4(1)=0
Inlist4 =0
Varlist5(1)=0
Inlist5 =0
'Open the input/output dialog box and select the analysis-variables,
'i.e., the dependent and independent variables for multiple regression
'analysis. Also check the options for the results to be displayed.
If SpecifyCaseSelectionCriterion = False Then Exit Sub
'Identify all possible combinations of the categories of the case-
'selection-variables and form strings, based on each combination of
'the categories, for the purpose of defining case selection
'conditions.
If VariableCategoryIdentifier( _
  NumberCriterionVars, MaxNoCriterionVars, _
  MaxNoCategories,NumberOfCategCombns) = _
          False Then Exit Sub
'Setup the workbook to include the results spreadsheets and plots.
Set ResultsWorkbook = Application.Workbooks.New
'Start the current analysis for all possible combinations of the
'categories of the case-selection-variables.
For i = 1 To NumberOfCategCombns
'Declare the Statistica module that is to be used for the current
'analysis as well as specify the data set to which the current
'analysis has to be restricted.
  Dim newanalysis As Analysis
  Set newanalysis = Analysis (scMultipleRegression, CurrentDataSet)
'Start opening and naming the folders in the workbook; each folder
'will include results for each combination of the categories of the
'case-selection-variables.
  Set Folder=ResultsWorkbook.InsertFolder( _
      ResultsWorkbook.Root, scWorkbookLastChild)
  Folder.Name=""
  Folder.Name= FolderTitle + FolderCaseName(i)
'If the case-selection-variable(s) have not been selected, then the
'analysis will proceed with the preset case selection conditions (if
'any) or else all cases will be considered for the current analysis.
  If(Inlist5=0) Then
    MsgBox "Case-selection variables have not been selected; " + _
      "cases with the preset selection conditions or all cases will " + _
```

CHAPTER 10: ADVANCED EXAMPLES

```
        "be used In the Analysis"
       GoTo PreConditionedAnalysis
      End If
      NewSelectionCond = AdditionalCaseSelectCond(i)
    PreConditionedAnalysis:
    'Concatenate the string of new set of case selection conditions with
    'the string of preset case selection conditions and supply the
    'resulting string, defining all case selection conditions, to the
    'current analysis.
      ifault =1
      Ret=AddSelectionCondition(NewSelectionCond, newanalysis)
      ifault=2
      If (Ret=99) Then GoTo kaputt
    'AUTOMATION INPUT SPECIFICATION SECTION
    'Supply automation input specifications.
    'Specify the analysis-variables, i.e., the variables involved in the
    'computations as well as the option for handling missing cases other
    'input options.
      If (Inlist1=0) Then
       MsgBox"Analysis-variables have not been selected; current " + _
         "analysis cannot be performed"
       GoTo kaputt:
      End If
      With newanalysis.Dialog
          .Variables =  Array(Varlist1, Varlist2)
          .InputFile = scRegRawData
          .CasewiseDeletionOfMD = True
          .PerformDefaultNonStepwiseAnalysis = False
          .ReviewDescriptiveStatistics = False
         .ExtendedPrecisionComputations = False
      End With
    'RUN the current analysis or proceed to process the next case, if
    'there is a problem with the current category combination or other
    'settings.
      On Error GoTo NextCase

      newanalysis.Run
    'AUTOMATION OUTPUT SPECIFICATION SECTION
    'Supply automation specifications for output (results).
      With newanalysis.Dialog
         .ComputeConfidenceLimits = True
         .AlphaForLimits = 0.050000
         .PLevelForHighlighting = 0.050000
      End With
    'RESULTS OPTION SECTION
    'First results option: Produce the results spreadsheet for summary of
```

112 – *STATISTICA* Visual Basic Primer

```vb
'regression results, using the current category combination as case
'selection criterion and include it as 'Summary: Regression results'
'in the current folder.
  If (ResultOption1="1") Then
    Set ResultsSpreadsheet = newanalysis.Dialog.Summary
    Set Spreadsheet = ResultsWorkbook.InsertObject( _
     ResultsSpreadsheet.Item(1), Folder, scWorkbookLastChild)
    Spreadsheet.Name ="Summary: Regression resuls"
  End If
'Second results option: Produce the results spreadsheet of Analysis of
'variance, using the current category combination as case selection
'criterion and include it as 'Analysis of variance' in the current
'folder.
  If (ResultOption2="1") Then
    Set ResultsSpreadsheet = newanalysis.Dialog.AnalysisOfVariance
    Set Spreadsheet = ResultsWorkbook.InsertObject( _
     ResultsSpreadsheet.Item(1), Folder, scWorkbookLastChild)
    Spreadsheet.Name ="Analysis of variance"
  End If
  newanalysis.Run
'Supply more specifications required for the computations of following
'results.
  With newanalysis.Dialog
      .RawResiduals = True
      .StandardResidualPlusMinusSigmaOutliers = True
      .RawPredictedValues = True
  End With
'Third results option: Produce the results spreadsheet for predicted
'and observed values, using the current category combination as case
'selection criterion and include it as 'Predicted and observed values'
'in the current folder.
  If (ResultOption3="1") Then
    Set ResultsSpreadsheet =newanalysis.Dialog.Summary
    Set Spreadsheet = ResultsWorkbook.InsertObject( _
     ResultsSpreadsheet.Item(1), Folder, scWorkbookLastChild)
    Spreadsheet.Name = "Predicted and observed values"
  End If
'Fourth results option: Produce the Normal probability plot of
'residuals, using the current category combination as case selection
'criterion and include it as 'Normal prob. plot of residuals' in the
'current folder.
  If (ResultOption4="1") Then
    Set ResGraphs=newanalysis.Dialog.NormalPlotOfResiduals
    Set Graph = ResultsWorkbook.InsertObject( ResGraphs.Item(1), _
     Folder, scWorkbookLastChild)
    Graph.Name = "Normal prob. plot of residuals"
```

```
        End If
        ResultsWorkbook.Visible = True
        GoTo Nexti
NextCase:
    MsgBox " Computations for " + Folder.Name + " failed."
Nexti:
    Next i       'Go to the next combination of categories of the case-
                 'selection-variable(s).
Exit Sub
'In case of error in the case selection conditions, display error
'message.
Kaputt:
If (ifault=1) Then
    MsgBox "No valid input file."
ElseIf (ifault=2) Then
    MsgBox "Cannot perform analysis-by-group, given the "+ _
           "case selection conditions in the current input file."
End If
'In case of any other error, terminate the job after displaying the
'error message.
MsgBox " Job is being terminated "
End Sub


'This function opens the input/output dialog box.
Function SpecifyCaseSelectionCriterion As Boolean
    SpecifyCaseSelectionCriterion=True
    Begin Dialog UserDialog 550,196, _
    "Multiple Regression Analysis by Variable(s)", _
    .DialogFunction
    ' %GRID:2,2,1,1
    OKButton 448,8,96,20,.OkButton
    CancelButton 448,32,96,20,.CancelButton
'Analysis-variables selection button:
    PushButton 8,8,84,20,"&Variables",.VariablesSelectionButton
    Text 196,56,80,18,"none",.List2Variables
    Text 196,34,80,18,"none",.List1Variables
'Case-selection-variables selection button:
    GroupBox 10,80,264,106,"Analysis by",.AnalysisOption
    PushButton 22,98,84,20,"V&ariable(s) : ",.CriterionButton
    GroupBox 286,80,256,106,"Results and plots",.GroupBox1
    CheckBox 300,98,231,16, _
      "Summary: Regression results",.CheckBox1
    CheckBox 300,120,231,16, _
      "Normal plot of residuals",.CheckBox2
    CheckBox 300,142,231,16, _
      "ANOVA (Overall goodness of fit)",.CheckBox3
```

```
    CheckBox 300,164,231,16, _
      "Predicted vs. Observed",.CheckBox4
    Text 22,34,172,18, _
      "Dependent variable",.Text2DependentVariables
    Text 22,56,172,18, _
      "Independent variables",.Text3IndependentVariables
    Text 24,126,240,56,"none",.Text1
    End Dialog
Dim dlg As UserDialog
Tryagain:
InitializeUserDialogFields (dlg)
SpecifyCaseSelectionCriterion=Dialog(dlg)
If SpecifyCaseSelectionCriterion = True Then
  If RetrieveUserDialogFields (dlg) =0 Then
    MsgBox "Error in the dialog entries; TRY Again.", vbCritical
    GoTo TryAgain
  End If
End If
End Function
'This function checks the validity of the preset case selection
'conditions, if any, and then retrieves them after concatenating
'with the new case selection conditions.
Private Function AddSelectionCondition (Condition  As String, _
  newanalysis As Analysis) As Integer
Dim InSelection As SelectionCondition
Dim MySelection As SelectionCondition
newanalysis.SelectionConditionSource=scSourceLocal
Set MySelection=newanalysis.SelectionCondition
'Initialize all preset case selection conditions.
MySelection.ExcludeExpression=""
MySelection.IncludeExpression=""
MySelection.Enabled=True
MySelection.ExcludeList=""
MySelection.IncludeList=""
'Check whether including the preset selection conditions will be
'feasible.
AddSelectionCondition=1
Set InSelection = CurrentDataSet.SelectionCondition
If (InSelection.Enabled And InSelection.IncludeList <> "") _
  Then GoTo kaputt
If (InSelection.Enabled) Then
  MySelection.ExcludeExpression=InSelection.ExcludeExpression
  MySelection.IncludeExpression=InSelection.IncludeExpression
  MySelection.ExcludeList=InSelection.ExcludeList
  MySelection.IncludeList=InSelection.IncludeList
End If
```

```
'If none of the strings defining the two types of selection
'conditions: new and preset are empty, then concatenate them.
'Otherwise, go ahead with the one which is not empty.
If (MySelection.IncludeExpression<>"" And Condition <> "") Then
 MySelection.IncludeExpression= "(" + _
  MySelection.IncludeExpression +")" + " and "+Condition
ElseIf (Condition ="") Then
 MySelection.IncludeExpression = MySelection.IncludeExpression
Else
 MySelection.IncludeExpression=Condition
End If
GoTo done
kaputt:      'In case of error.
AddSelectionCondition=99
done:
End Function
'This subroutine initializes the necessary entries in the input/output
'dialog box.
Sub InitializeUserDialogFields (dlg)
'Initially check all the options for the results to be displayed.
 dlg.CheckBox1 =1
 dlg.CheckBox2 =1
 dlg.CheckBox3 =1
 dlg.CheckBox4 =1
End Sub
'This function retrieves the selected entries from the dialog box.
Function RetrieveUserDialogFields (dlg) As Boolean
 On Error GoTo InvalidInput
 RetrieveUserDialogFields=True
'Retrieve the states of check boxes that have finally been checked by
'the user.
 ResultOption1 = Trim$(Str(dlg.CheckBox1))
 ResultOption2 = Trim$(Str(dlg.CheckBox2))
 ResultOption3 = Trim$(Str(dlg.CheckBox3))
 ResultOption4 = Trim$(Str(dlg.CheckBox4))
 Exit Function
 InvalidInput:
 RetrieveUserDialogFields=False
End Function
'This private function controls the functionality of different tabs in
'the input/output dialog box.
Private Function DialogFunction(DlgItem$, Action%, _
 SuppValue&) As Boolean
Dim AnalysisVarName As String
Select Case Action%
 Case 1
```

```
'Dialog box initialization
 Case 2
'Value changing or button pressed
 DialogFunction = True
 Select Case DlgItem
'Control for selection of the case-selection-variable(s)
   Case "CriterionButton"
'Select the case-selection-variable(s), using one-variable selection
'list.
   SelectVariableForAnalysis(Varlist5, Inlist5)
   If (Inlist5=0) Then GoTo AnalysisVarSelection
   ReDim VarCodeNumber(Inlist5)
   For i =1 To Inlist5
    VarCodeNumber(i) = CInt(Varlist5(i))
   Next i
'Pick up the names of the selected case-selection variables. These
'names will be displayed on the input/output dialog box after the
'variables have been selected.
   AnalysisVarName = ""
   For j=1 To Inlist5
    If (Inlist5 > 10) Then
     If (j=Inlist5) _
     Then AnalysisVarName =  AnalysisVarName  + _
      Str$(VarCodeNumber(j))
     Else
      AnalysisVarName =  AnalysisVarName  + _
       Str$(VarCodeNumber(j)) + ", "
     End If
    If (j=Inlist5) Then
     AnalysisVarName =  AnalysisVarName  + _
      CurrentDataSet.VariableName(VarCodeNumber(j))
    Else
     AnalysisVarName =  AnalysisVarName  + _
      CurrentDataSet.VariableName(VarCodeNumber(j))+ ", "
    End If
   Next j
   DlgText "Text1" , AnalysisVarName
   DialogFunction=True
   AnalysisVarSelection:
'Control for the selection of analysis-variables:
   Case "VariablesSelectionButton"
'Select the analysis-variables. The following function will display as
'many lists as the number of analysis-variables lists, i.e.,
''NumberAnalysisVarLists', supplied by the user.
   VariablesSpecifications(Varlist1,Varlist2, Varlist3, Varlist4 )
   DialogFunction=True
```

CHAPTER 10: ADVANCED EXAMPLES

```
'The button that triggers the computations of the results for the
'current input/output settings.
    Case "OkButton"
    DialogFunction=False
'The button that cancels the job.
    Case "CancelButton"
    MsgBox "Job is being cancelled by the user"
    DialogFunction=False
  End Select
End Select
End Function

'This function selects the case-selection-variable(s).
Function SelectVariableForAnalysis(Varlist5, Inlist5) As Boolean
On Error GoTo ErrorMessage
SelectVariableForAnalysis=True
Dim Nvars As Long
Dim Ret As Integer
Ret=1
Nvars = CurrentDataSet.NumberOfVariables
'Select the case-selection-variable(s) from the one-variable list.
Ret = SelectVariables1 ( CurrentDataSet, _
  "Select grouping variable(s) for analysis", 0, Nvars, _
  Varlist5, Inlist5, "Case-selection-variables")
NumberCriterionVars = Inlist5
Exit Function
ErrorMessage:
MsgBox " Error in the selection of variable for analysis"
SelectVariableForAnalysis=False
End Function

'This function (i) identifies all possible combinations of the
'categories of the case-selection-variables, (ii) creates the strings
'defining the new case selection conditions, and (iii) provides the
'names (using these strings) of the folders that will contain the
'results for each combination of the categories of the case-
'selection-variables.
Function VariableCategoryIdentifier (NumberCriterionVars , _
  MaxNoCriterionVars , MaxNoCategories , _
  CountCatComb As Integer) As Boolean
On Error GoTo ErrorMessage
VariableCategoryIdentifier=True
ReDim VariableCateg(1 To MaxNoCategories, _
  1 To MaxNoCriterionVars) As String
ReDim VarCategCode(1 To MaxNoCategories, _
  1 To MaxNoCriterionVars) As Long
```

CHAPTER 10: ADVANCED EXAMPLES

```
Dim i As Long, j As Long
Dim nr As Long, nc As Long
Dim vr As Integer
Dim CategCombination As String
nr=CurrentDataSet.NumberOfCases
nc=CurrentDataSet.NumberOfVariables
ReDim r(nr) As Range
CountCatComb =1
'Keep checking the cells in the columns corresponding to the case-
'selection-variable(s) through the rows, till a row (case) is located
'in which NONE of the cells is empty. This gives the first set of
'valid cells for the purpose of category identification. At that
'point get out of the loop for the cases.
For i=1 To nr
  For j=1 To NumberCriterionVars
    If CurrentDataSet.MissingData(i,VarCodeNumber(j)) Then
      GoTo Nexti
    Else
      Set r(j) = CurrentDataSet.Cells (i,VarCodeNumber(j))
    End If
  Next j
  GoTo OutOfLoop
Nexti:
Next i
OutOfLoop:
'Retrieve the texts available in the valid cells located above. This
'will provide the first combination of the categories of the case-
'selection-variables.
For j=1 To NumberCriterionVars
  VariableCateg(j,CountCatComb) = r(j).Text
  VarCategCode(j,CountCatComb) = r(j)
Next j
'Compare the currently identified combination of the categories with
'the entries in the cells of the selected columns for each row in
'order to identify the next possible combination. As the comparison in
'the row containing an empty cell for the selected columns will not
'make sense, each row in the selected columns have to be scanned for
'empty cell(s) and a row has to be skipped even if one of the cells in
'the selected columns is empty.
For vr = 1 To nr
  For j=1 To NumberCriterionVars
    If CurrentDataSet.MissingData(vr,VarCodeNumber(j)) Then
      GoTo NextCase
    Else
      Set r(j) = CurrentDataSet.Cells (vr,VarCodeNumber(j))
    End If
```

Chapter 10: Advanced Examples

```
    Next j
'If all the cells for the selected columns in the current row are non-
'empty, then compare the texts available in these cells with all
'previously identified combinations.
    For i= 1 To CountCatComb
      For j=1 To NumberCriterionVars
        If (r(j).Text = VariableCateg(j,i)) Then
          GoTo jlist
        Else
'If one of the text entries in the current row form a combination that
'is different from all the previously identified combinations, then
'this is the next possible combination of the categories that is being
'searched.
          If(i= CountCatComb) Then
            GoTo AddCount
          Else
            GoTo iCount
          End If
        End If
        jlist:
      Next j
    GoTo NextCase
iCount:
    Next i
AddCount:
'Increment the number of identified categories as well as save the
'combination to the list of combinations of the categories that have
'already been identified.
    CountCatComb = CountCatComb +1
    For j=1 To NumberCriterionVars
      VariableCateg(j,CountCatComb) = r(j).Text
      VarCategCode(j,CountCatComb) = r(j)
    Next j
    NextCase:
Next vr
'After all the combinations have been identified, create strings
'defining the new sets of case selection conditions. These will
'later be concatenated with the preset case selection conditions.
ReDim AdditionalCaseSelectCond(1 To CountCatComb)
ReDim   FolderCaseName (1 To CountCatComb) As String
'Start concatenating the strings defining the category combinations.

For i=1 To CountCatComb          'For each identified category
  'combination
  AdditionalCaseSelectCond(i)=""
  For j = 1 To Inlist5                    'For each case-selection variable
```

120 – *STATISTICA* Visual Basic Primer

```
'Create the string that will define the i-th combination of the
'categories of the j-th case-selection-variable, i.e., the string of
'the type, "vj=i".
  CriterionVariable = vbNullChar
  CriterionVariable = Trim$(Str$(Varlist5(j)))
  If (j<Inlist5) Then
    AdditionalCaseSelectCond(i) = AdditionalCaseSelectCond(i) + _
       "v" + CriterionVariable + "=" + Trim$(Str$(VarCategCode(j,i))) _
       + " and "
  Else
    AdditionalCaseSelectCond(i) = AdditionalCaseSelectCond(i) + _
       "v" + CriterionVariable + "=" + Trim$(Str$(VarCategCode(j,i)))
  End If
 Next j
'Create the string containing the name of the i-th folder, i.e., the
'folder that will contain the results for the i-th category
'combination used in the case selection.
 If(Inlist5=0) Then
  FolderCaseName(i)=""
 Else
  FolderCaseName(i)=""
  CategCombination=""
  For j=1 To NumberCriterionVars       'For each case-selection-variable
    If j=NumberCriterionVars Then
      CategCombination = CategCombination + VariableCateg(j,i)
    Else
      CategCombination = CategCombination + VariableCateg(j,i) + ", "
    End If
  Next j
  FolderCaseName(i) =  " by: " + CategCombination
 End If
Next i         'Go to the next combination of categories
Exit Function
ErrorMessage:
MsgBox " Error in the identification of the code " + _
   "for analysis_variable"
VariableCategoryIdentifier=False
End Function
'This function selects the analysis-variables, from one-, two-, three-
'or four-variables lists, depending upon the number of analysis-
'variables lists, i.e., 'NumberAnalysisVarLists' supplied by the user
'for the current analysis. This function also confirms if the
'variable(s) from the lists(s) have been selected, by displaying the
'status as 'Selected' in the dialog box.
Function VariablesSpecifications(Varlist1() As Integer, _
  Varlist2() As Integer, Varlist3() As Integer, _
```

CHAPTER 10: ADVANCED EXAMPLES

```
    Varlist4() As Integer )   As Boolean
On Error GoTo ErrorMessage
VariablesSpecifications=True
Dim i As Integer
Dim Ret As Integer
'Decide to use one-, two-, three-, or four-variables lists, depending
'upon the value of 'NumberAnalysisVarLists'.
Select Case NumberAnalysisVarLists
Case "1"     'One-variable list selection
Ret = SelectVariables1(CurrentDataSet, _
 "Variables for Analysis", _
 0, Nvars, Varlist1, Inlist1, List1Title)
 If Ret=0 Then
  VariablesSpecifications=False
 Else
  VariablesSpecifications=True
 End If

Case "2"     'Two-variables lists selection
Ret = SelectVariables2(CurrentDataSet, _
 "Variables for Analysis", _
 0, Nvars, Varlist1, Inlist1, List1Title, _
 0, Nvars, Varlist2, Inlist2, List2Title)
VariablesSpecifications=True

Case "3"     'Three-variables lists selection
Ret = SelectVariables3(CurrentDataSet, _
 "Variables for Analysis", _
 0, Nvars, Varlist1, Inlist1, List1Title, _
 0, Nvars, Varlist2, Inlist2, List2Title, _
 0, Nvars, Varlist3, Inlist3, List3Title)
VariablesSpecifications=True

Case "4"     'Four-variables lists selection
Ret = SelectVariables4(CurrentDataSet, _
 "Variables for Analysis", _
  0, Nvars, Varlist1, Inlist1, List1Title, _
  0, Nvars, Varlist2, Inlist2, List2Title, _
  0, Nvars, Varlist3, Inlist3, List3Title, _
  0, Nvars, Varlist4, Inlist4, List4Title)
VariablesSpecifications=True
End Select
'Confirm the selection of analysis-variables, and, if selected,
'display as 'Selected' in the input/output dialog box.
VariablesSelectionConfirmation
Exit Function
ErrorMessage:
```

```
VariablesSpecifications=False
MsgBox " Error in the selection of variables involved in computations"
End Function
'This function confirms whether or not the variables in each list of
'the analysis-variables have been selected.
Function VariablesSelectionConfirmation As Boolean
'If the List1 variables have been selected, then mark them
'Selected' 'in the input/output dialog box, otherwise unselect
' all and mark them 'none'.
If (CInt(NumberAnalysisVarLists) > 0) Then
 If(Inlist1 >0) Then
  DlgText "List1Variables" , "Selected"
 Else
  DlgText "List1Variables" , "none"
  For i = 1 To Nvars
   Varlist1(i)= 0
  Next i
 End If
End If
'If the List2 variables have been selected, Then mark them 'Selected'
'in the input/output dialog box, otherwise unselect all and mark them
''none'.
If (CInt(NumberAnalysisVarLists) > 1) Then
 If(Inlist2 >0) Then
  DlgText "List2Variables" , "Selected"
 Else
  DlgText "List2Variables" , "none"
  For i = 1 To Nvars
   Varlist2(i)= 0
  Next i
  End If
 End If
'If the List3 variables have been selected, then mark them 'Selected'
'in the input/output dialog box, otherwise unselect all and mark them
''none'.
If (CInt(NumberAnalysisVarLists) > 2) Then
 If(Inlist3 >0) Then
  DlgText "List3Variables" , "Selected"
 Else
  DlgText "List3Variables" , "none"
  For i = 1 To Nvars
   Varlist3(i)= 0
  Next i
 End If
End If
'If the List4 variables have been selected, then mark them 'Selected'
```

CHAPTER 10: ADVANCED EXAMPLES

```
'in the input/output dialog box, otherwise unselect all and mark them
''none'.
If (CInt(NumberAnalysisVarLists) > 3) Then
 If(Inlist4 >0) Then
  DlgText "List4Variables" , "Selected"
 Else
  DlgText "List4Variables" , "none"
  For i = 1 To Nvars
   Varlist4(i)= 0
  Next i
 End If
End If
End Function
```

124 – *STATISTICA* Visual Basic Primer

A
APPENDIX

DOCUMENT-LEVEL EVENT COMMANDS

Workbook Events 127
Analysis Events 128
Spreadsheet Events 128
Report Events 130
Graph Events 131

APPENDIX A

DOCUMENT-LEVEL EVENT COMMANDS

Workbook Events

The following table includes the document-level events that are available for workbooks. For more information on events, see *Controlling STATISTICA Events with SVB Programs* (page 76).

Command	Action
Activate	Executes when the workbook receives the focus within the *STATISTICA* workspace (typically when you click on the workbook or a document is sent to it).
BeforeClose	Executes when you close the workbook. Before the document closes, the events within this function will first execute. An example of using this feature would be to add additional functionality, such as saving the file to multiple locations or exporting it to another application.
BeforePrint	Executes when you print the workbook. Before the workbook prints, the events within this function will first execute. Examples of using this feature would be to prevent users from printing the document or adding new functionality such as calling an alternate, customized printing dialog.
BeforeRightClick	Executes when you right-click anywhere on the workbook. Before the shortcut menu is displayed, the events within this function will first execute. An example of using this feature would be to add additional functionality to displaying shortcut menus, such as selecting the entire contents of the workbook.

APPENDIX A: DOCUMENT-LEVEL EVENT COMMANDS

BeforeSave	Executes when you save changes made to the workbook. Before the changes are actually saved, the events in this function will first execute. An example of using this feature would be to prevent users from permanently changing a workbook.
Deactivate	Executes when the workbook has lost the focus to another window within *STATISTICA*.
OnClose	Executes when the workbook is being closed.
Open	Executes when the workbook is being opened. Note, this will not run unless autorun was specified when the macro was last edited.
SelectionChanged	Executes when the focus within the workbook has moved.

Analysis Events

The following table includes the available analysis-level events. For more information on events, see *Controlling STATISTICA Events with SVB Programs* (page 76).

Command	Action
Activate	Executes when the analysis receives the focus within the *STATISTICA* workspace.
BeforeClose	Executes when you close the analysis. Before the analysis closes, the events within this function will first execute.
BeforeOutput	Executes when an analysis within *STATISTICA* is about to send output to a workbook, spreadsheet, etc.; a typical application using this event is to implement custom-handling of output, for example, to send all graphs and results spreadsheets to another application.
Deactivate	Executes when the analysis has lost the focus to another window within *STATISTICA*.
OnClose	Executes when the analysis is being closed.

Spreadsheet Events

The following table includes the document-level events that are available for spreadsheets. For more information on events, see *Controlling STATISTICA Events with SVB Programs* (page 76).

APPENDIX A: DOCUMENT-LEVEL EVENT COMMANDS

Command	Action
Activate	Executes when the spreadsheet receives the focus within the *STATISTICA* workspace (typically, when you click on the spreadsheet).
BeforeClose	Executes when you close the spreadsheet. Before the document closes, the events within this function will first execute. An example of using this feature would be to add additional functionality, such as saving the file to multiple locations or exporting it to another application.
BeforeDoubleClick	Executes when you double-click anywhere on the spreadsheet. Before the clicked area goes into edit mode, the events within this function will first execute. Examples of using this feature would be to prevent users from altering a spreadsheet's content or to add additional functionality such as applying a custom format to the clicked area.
BeforePrint	Executes when you select a print option. Before the spreadsheet prints, the events within this function will first execute. Examples of using this feature would be to prevent users from printing the document or adding new functionality such as calling an alternate, customized printing dialog.
BeforeRightClick	Executes when you right-click anywhere on the spreadsheet. Before the shortcut menu is displayed, the events within this function will first execute. An example of using this feature would be to add additional functionality to displaying shortcut menus, such as highlighting the entire contents of the spreadsheet.
BeforeSave	Executes when you save changes made to the spreadsheet. Before the changes are actually saved, the events in this function will first execute. An example of using this feature would be to prevent users from permanently changing a spreadsheet.
DataChanged	Executes when data within a cell or block of cells is altered. This will also execute when a variable's Data type or Display format is changed.
Deactivate	Executes when the spreadsheet has lost the focus to another window within *STATISTICA*.
OnClose	Executes when the spreadsheet is being closed.
Open	Executes when the spreadsheet is being opened. Note that this will not run unless autorun was specified when the macro was last edited.

APPENDIX A: DOCUMENT-LEVEL EVENT COMMANDS

SelectionChange — Executes when the focus within the spreadsheet has moved.

StructureChanged — Executes when the size of the spreadsheet is changed. Examples of events that trigger this include the addition or deletion of variables or cases.

Report Events

The following table includes the document-level events that are available for reports. For more information on events, see *Controlling STATISTICA Events with SVB Programs* (page 76).

Command	Action
Activate	Executes when the report receives the focus within the *STATISTICA* workspace (typically when you click on the report or a document is sent to it).
BeforeClose	Executes when you close the report. Before the document closes, the events within this function will first execute. An example of using this feature would be to add additional functionality, such as saving the file to multiple locations or exporting it to another application.
BeforePrint	Executes when you print a report. Before the report prints, the events within this function will first execute. Examples of using this feature would be to prevent users from printing the document or adding new functionality such as calling an alternate, customized printing dialog.
BeforeRightClick	Executes when you right-click anywhere on the report. Before the shortcut menu is displayed, the events within this function will first execute. An example of using this feature would be to add additional functionality to displaying shortcut menus, such as highlighting the entire contents of the report.
BeforeSave	Executes when you save changes made to the report. Before the changes are actually saved, the events in this function will first execute. An example of using this feature would be to prevent users from permanently changing a report.
Deactivate	Executes when the report has lost the focus to another window within *STATISTICA*.
OnClose	Executes when the report is being closed.

Open	Executes when the report is being opened. Note, this will not run unless autorun was specified when the macro was last edited.
SelectionChanged	Executes when the focus within the report has moved.

Graph Events

The following table includes the document-level events that are available for graphs. For more information on events, see *Controlling STATISTICA Events with SVB Programs* (page 76).

Command	Action
Activate	Executes when the graph receives the focus within the *STATISTICA* workspace.
BeforeClose	Executes when you close the graph. Before the document closes, the events within this function will first execute. An example of using this feature would be to add additional functionality, such as saving the file to multiple locations or exporting it to another application.
BeforeDoubleClick	Executes when you double-click anywhere on the graph. Before an edit dialog is displayed (depending on which area of the graph was double-clicked on), the events within this function will first execute. Examples of using this feature would be to prevent users from altering a graph's content or to add additional functionality such as applying a custom style to the clicked area.
BeforePrint	Executes when you print a graph. Before the graph prints, the events within this function will first execute. Examples of using this feature would be to prevent users from printing the document or adding new functionality such as calling an alternate, customized printing dialog.
BeforeRightClick	Executes when you right-click anywhere on the graph. Before the shortcut menu is displayed, the events within this function will first execute. An example of using this feature would be to add additional functionality to displaying shortcut menus, such as highlighting the entire contents of the graph.
BeforeSave	Executes when you save changes made to the graph. Before the changes are actually saved, the events in this function will first execute. An example of using this feature would be to prevent users from permanently changing a graph.

APPENDIX A: DOCUMENT-LEVEL EVENT COMMANDS

Deactivate — Executes when the graph has lost the focus to another window within *STATISTICA*.

OnClose — Executes when the graph is being closed.

Open — Executes when the graph is being opened. Note, this will not run unless autorun was specified when the macro was last edited.

B
APPENDIX

APPLICATION-LEVEL EVENT COMMANDS

APPENDIX B

APPLICATION-LEVEL EVENT COMMANDS

The following table gives the application-level events that are available in *STATISTICA*. For more information on events, see *Controlling STATISTICA Events with SVB Programs* (page 76).

Command	Action
AnalysisActivate	Executes when a *STATISTICA* analysis within the *STATISTICA* environment has received the focus.
AnalysisBeforeClose	Executes when a *STATISTICA* analysis within *STATISTICA* is about to be closed.
AnalysisBeforeOutput	Executes when an analysis within *STATISTICA* is about to be processed.
AnalysisDeactivate	Executes when an analysis has lost the focus to another window within *STATISTICA*.
AnalysisNew	Executes when *STATISTICA* creates a new analysis.
AnalysisOnClose	Executes when *STATISTICA* closes an analysis.
OnActiveDataSet	Executes whenever the program makes reference to the `ActiveDataSet`; this event handler would, for example, be useful to automatically replace in an SVB program references to an `ActiveDataSet` with a custom routine.
OnClose	Executes when *STATISTICA* is being closed.
OnInit	Executes when *STATISTICA* is opening (initialized).

APPENDIX B: APPLICATION-LEVEL EVENT COMMANDS

ReportActivate	Executes when a *STATISTICA* Report within the *STATISTICA* environment has received the focus.
ReportBeforePrint	Executes when you print a report within *STATISTICA*.
ReportBeforeRightClick	Executes when you right-click anywhere on a report within *STATISTICA*.
ReportBeforeSave	Executes when you save changes made to the report within *STATISTICA*.
ReportDeactivate	Executes when a report has lost the focus to another window within *STATISTICA*.
ReportNew	Executes when *STATISTICA* creates a new report.
ReportOnClose	Executes when *STATISTICA* closes a report.
ReportOpen	Executes when *STATISTICA* opens a report.
ReportSelectionChanged	Executes when the focus within a report in *STATISTICA* has moved.
SpreadsheetActivate	Executes when a *STATISTICA* Spreadsheet within *STATISTICA* has received the focus.
SpreadsheetBeforeClose	Executes when a *STATISTICA* Spreadsheet within *STATISTICA* is about to be closed.
SpreadsheetBeforeDoubleClick	Executes when you double-click anywhere on a spreadsheet within *STATISTICA*.
SpreadsheetBeforePrint	Executes when you print a spreadsheet within *STATISTICA*.
SpreadsheetBeforeRightClick	Executes when you right-click anywhere on the spreadsheet within *STATISTICA*.
SpreadsheetBeforeSave	Executes when you save changes made to the spreadsheet within *STATISTICA*.
SpreadsheetDataChanged	Executes when data within a cell or block of cells of a spreadsheet within *STATISTICA* has been altered. This will also execute when a variable's Data type or Display format has changed.

APPENDIX B: APPLICATION-LEVEL EVENTS COMMANDS

SpreadsheetDeactivate	Executes when a spreadsheet has lost the focus to another window within *STATISTICA*.
SpreadsheetNew	Executes when *STATISTICA* creates a new spreadsheet.
SpreadsheetOnClose	Executes when *STATISTICA* closes a spreadsheet.
SpreadsheetOpen	Executes when *STATISTICA* opens a spreadsheet.
SpreadsheetSelectionChange	Executes when the focus within a spreadsheet in *STATISTICA* has moved.
SpreadsheetStructureChanged	Executes when the size of a spreadsheet within *STATISTICA* is changed.
WorkbookActivate	Executes when a *STATISTICA* Workbook within the *STATISTICA* environment has received the focus.
WorkbookBeforeClose	Executes when you close a workbook in *STATISTICA*.
WorkbookBeforePrint	Executes when you print a workbook in *STATISTICA*.
WorkbookBeforeRightClick	Executes when you right-click anywhere on a workbook in *STATISTICA*.
WorkbookBeforeSave	Executes when you save changes made to a workbook in *STATISTICA*.
WorkbookDeactivate	Executes when a workbook in *STATISTICA* has lost the focus to another window.
WorkbookNew	Executes when *STATISTICA* creates a new workbook.
WorkbookOnClose	Executes when a workbook in *STATISTICA* is being closed.
WorkbookOpen	Executes when *STATISTICA* opens a workbook.
WorkbookSelectionChanged	Executes when the focus within a workbook in *STATISTICA* has moved.

INDEX

A

accessing SVB libraries, 50
advanced examples, 97
analysis
    by group, 107
    -level event commands, 128
analysis macros, 11, 13
    case selection conditions, 14
    case weights, 14
    datafile selections and operations, 14
    example, 15
    output options, 15
    results spreadsheets and graphs, 15
analysisoutput objects, 63
application-level event commands, 135
applications for SVB, 5
arrays, 31
automatic programming, 22

B

basic rules for simple SVB programs, 30
byref (by reference), passing arguments, 32
byval (by value), passing arguments, 33

C

case selection conditions, 107
    recording in master macros (logs), 21
case weights, recording in master macros (logs), 21
cell-function spreadsheets, 104
check boxes, 58
codes, specifying in SVB, 56, 57
collections, 32, 41, 63
    retrieving, 42
combo boxes, 59
common elements, 55
consecutive analyses, 20
creating an SVB program, 3
creating output documents, 41, 90, 91, 93
custom dialogs, 67
    defining dialogs in subroutines, 75
    examples, 67
    retrieving numeric values, 75
    servicing controls, 71, 72
customizing
    graphs, 98
    menus, 79
    toolbars, 79

D

data types, 30
debugging SVB programs, 7, 63
dialogs, implied in the program, 40
DLLs, external
    calling functions, 36
document-level event commands, 127
    graph, 131
    report, 130
    spreadsheet, 128
    workbook, 127

E

editor, for SVB programs, 7
events
    analysis-level events, 128
events (cont.)
    application-level events, 76, 135
    controlling, 76
    document-level events, 76, 127
    example SVB program, 77, 78, 104
    types of events, 76
executing an SVB program, 4
external DLLs
    calling functions, 36

F

functions, 30

G

global
    macro, 24
    variables, 33
goback method, 40, 60
graph objects, 98

I

if...then...end if block, 29
introductory examples, 89

K

keyboard macros, 11, 13

L

libraries, 47
list boxes, 59
logs of analyses, 12, 17

M

master macros (logs), 11, 12, 17
    adding, moving, copying, and deleting variables and cases, 19

master macros (logs) (cont.)
    ambiguities in recording and running macros, 20
    analysis macro recording, 21
    applications, 22
    case selection conditions, 20, 21
    case weights, 20, 21
    consecutive analyses, 20
    creating subsets of cases or variables, 19
    data editing operations, 19
    data transformations, 19
    datafile selections and operations, 18
    date operations, 19
    missing data replacement, 19
    opening and saving datafiles, 19
    output options, 21
    ranking data, 19
    recoding operations, 19
    results spreadsheets and graphs, 21
    shifting of data columns, 19
    simultaneous analyses, 20
    sorting of data, 19
    standardizing data, 19
    variable specifications, 19
matrix functions, 83
    example, 84
    STB.svx file, 83
message box, 29
methods, 34
modules and libraries, 47
moving between dialogs, 40

N

numeric input fields, 58

O

object model, 39, 40, 55
objects, 34
option buttons, 58
organization of programs, 40
output
    documents, 41, 43, 63
    objects, 41, 43, 63
output documents, 41
overview, 3

P

passing
    arguments to functions, 32
    array arguments, 33
performing computations, 30
programming
    dialogs, 40
    environment, 29
progress bar, creating in SVB, 90
properties, 34

R

recording
    an analysis macro, 22
    an entire analysis, 11, 12, 17
    analysis macros, 11, 13
    case selection conditions, 12, 14
    case weights, 12, 14
    keyboard macros, 11, 13
    macros, 11
    master macros, 11, 12, 17
    output options, 15, 21
    overview, 11
    results spreadsheets and graphs, 15, 21
reference libraries, 47
results
    graphs, 41, 63
    selections, 61

results (cont.)
    spreadsheets, 41, 63
    values, 62
    variables, 61
routeoutput method, 41, 43, 63
run method, 40, 60

S

specifying
    variable lists, 56
    variables, 56
statistical functions, 83
    example, 84
    STB.svx file, 83
structure of SVB, 5
subroutines, 30
symbolic constants, 59

T

type-ahead help, 7

V

variable lists, 56
variables, 56
variant data type, 32

W

with ... end with block, 56
write protect cells, 106